MADE IN OHIO

MADE IN OHIO

A HISTORY OF BUCKEYE INVENTION & INGENUITY

CONRADE C. HINDS

Published by The History Press
Charleston, SC
www.historypress.com

Front cover, top left: courtesy Hasbro Inc.; *top right*: courtesy of Ideal Electric Company; *bottom left*: courtesy Columbus Washboard Company Collection; *bottom right*: courtesy Ohio Art Company.
Back cover, top: courtesy Sutphen Corporation; *bottom*: courtesy of J.M. Smucker Company.

First published 2023

Manufactured in the United States

ISBN 9781467152945

Library of Congress Control Number: 2022947993

This work is dedicated to my grandparents the late Ethel and Russell Davidson, who were employed in Cleveland factories; and to sisters Patricia and Sandra, who were born in the industrious community of Xenia, Ohio; and also to my granddaughter Esmeralda Hinds.

CONTENTS

PREFACE

Unfortunately, too many Ohioans have forgotten or are not aware of the state's heritage of applying ingenuity toward inventing and manufacturing. Invention is the result of technically addressing a need that generally starts by presenting itself as a problem. Once the inventor has triumphed with a solution, the next goal is to find or create a demand. Manufacturing then proceeds to fill that demand. This then translates into employment for skilled and unskilled workers and business for suppliers and retailers.

Regretfully, this book is just a sampling of the great minds of male and female inventors and problem solvers who worked to advance the welfare of society in general. The list of large and small manufacturers that have come, gone and are now forgotten would read like a vast cemetery of unmarked graves called "Rustbelt Acres." But I think there is still a benefit and much to be learned by properly acknowledging and embracing our heritage of ingenious productivity. Too often, entertainment and the news media have the public's attention focused on unproductive pleasures, excitement, violence and disaster, and for many, this is an education on how to waste precious time. Some people have adopted the mindset that not working is their job, while many businesses are challenged in hiring capable employees.

Five of the biggest issues today that present endless opportunities for innovative problem solving and employment are rehabilitating twentieth-century infrastructure, climate change, energy, environmental sustainability and recycling resources such as water. These are issues of need, just like the nineteenth-century's need for roads, bridges, railroads, clean drinking water, electricity and better light. These issues were tackled in the wake of a wasteful and devastating civil war fought over trying to preserve the cruel and inefficient institution of slavery. This millennium started with a world population of six billion; in less than twenty-three years, we are approaching eight billion people. They will need to be housed, clothed and fed with the limited resources available on earth. In this I see an ocean full of opportunities and a need for Ohio-style inventiveness because as a global society we have a lot of work ahead of us.

ACKNOWLEDGEMENTS

Thank you to Tom Betti and Doreen Uhas Sauer for getting me started as an author of lost and forgotten history. Special thanks to Josh and Eleanor Walters for the use of the pleasant environment of their home to get this project off the ground. Also, to my late parents and grandparents, Dr. Conrade and Ada Hinds, and Russell and Ethel Davidson, respectively, for sharing their insight and wisdom.

I'd like to extend my warm acknowledgements to Bradley Belden and Kristie Fitzgerald of Belden Brick Company, Mark Smucker and Frank Cirillo of J.M. Smucker Company, Bruno Maier and Kris Parlett of United Wheels Inc., Jacqui Barnett of Columbus Washboard Company, Elena West and Bill Killgallon of Ohio Art Company, Drew Sutphen and Josh Plichta of Sutphen Corporation, Abby Volk of Ideal Electric Company, and Alex Bandar of Columbus Idea Foundry for their valuable contributions. And my continued thanks to all those who have supported preserving Ohio's great heritage of invention, innovation and manufacturing.

A special note of gratitude to the many people of the "Greatest Generation" who are no longer with us for sharing their wisdom and positive encouragement and opening the doors of opportunity for me during the latter twentieth century.

INTRODUCTION

THE STORY OF OHIO'S INVENTIONS AND MANUFACTURING IS A HISTORY OF WHAT OHIO DOES BEST

The United States has lost an estimated 6.7 million manufacturing jobs in four decades due to decreased output resulting from global competition. Global competition is something that many Americans don't understand and don't want to face because of their national pridefulness. That pridefulness is a barrier to understanding what made the country so strong, especially in the wake of World War II. Those of the baby boom generation and beyond miss the point that in 1946, when the baby boom generation began, America in most cases had no global competition, a result of the devastation of the developed world during global war. England was still rationing food well into the 1950s. Meanwhile, in America, after a brief recession while the economy shifted from wartime to peacetime production, the manufacturing sector ramped up at full speed to meet the great demand for homes, cars, appliances and a host of other consumer goods. The country's population grew from 140 million in 1946 to 218 million in 1976. That's twice the growth from the previous thirty years starting in 1916. The America of the 1950s through the 1970s was a country of unprecedented economic growth and expansion.

Ohio had been a strong player in the production of manufactured goods since the mid-1850s. Much of this has to do with the innovative genius of many rural men and women who were just plain fed up with the drudgery of agricultural tasks requiring hand tools and general manual labor. I personally would like to have the state recognized as the nation's official "mother of invention" because of our heritage of applied innovation to solve problems. Once a problem is solved via a practical invention, the product then has to be manufactured in quantity and distributed to members of the public with the same problem. This creates a market demand.

Manufacturing to meet that demand requires planning; acquiring raw materials, a source of energy, machinery, tooling and equipment; hiring skilled—and training unskilled—labor; conducting quality control; packaging; advertising; and transportation. In twentieth-century Ohio, there was some kind of tool-and-die or machine shop every few blocks in most sizable communities. Many of those businesses have been replaced by tattoo parlors and computer/cell phone repair stores and coffee shops.

But I believe the spirit of innovation is still alive in Ohio. But the world's population is exponentially increasing, and we are adding to the issue of polluting our land, air and oceans. Cleaning the planet is a task that requires more than just a manufactured gadget or device. The innovation needed here is a way to increase and maximize cooperation, education and mutual appreciation of the problems that future generations must face and solve or adapt to in order to live a life with a respectable standard of living and well-being.

The skylines of established, healthy American communities were first defined by church steeples, in the days when much of the country was an agrarian-based society. With the emergence of the industrial economy and the steam age, the church steeples were overtaken in height by factory smokestacks. Burning coal emitted a cloud of black smoke that covered the community, and in general there were no complaints. That's because a factory's chimney smoke meant employment and prosperity. But in time, even the smokestack, because of environmental regulations, was replaced by another international symbol of a strong economy: the skyscraper.

The following chapters will look back at the innovations and manufactured goods of Ohio that shaped not only America but also the world. Manufacturing in Ohio has been at the forefront of producing building materials, glassware, furniture, household appliances and even complex high-technology items such as automobiles, aircraft and military hardware such as tanks. Equipment for mining raw materials and excavating for mass infrastructure projects have also been fabricated or manufactured in the state. One testament to the excellent job Ohioans have done over the past century is that Americans became so accustomed to having high-quality goods manufactured in Ohio that they began to take many things for granted. One example is the mid-twentieth-century

Frigidaire refrigerator, assembled in Dayton, Ohio. It could easily last a generation without any need for maintenance.

I trust that this book will serve to remind Ohioans of their great manufacturing heritage while opening young eyes to the even greater challenges that face us in the present and subsequent future.

1

RESHAPING THE NORTHWEST TERRITORY OHIO LAND

For several decades, there has been talk of the former glory days of the midwestern industrial Rust Belt region of America and the undermining of the middle class. It's unfortunate that many have no idea of the major part that Ohio played in developing this great nation and that the state is still very active in helping to sustain the well-being of our country. There is a slogan frequently used that states, "Ohio, the Heart of It All." But in reality, I like to think of Ohio as being where the heartbeat of America was started. The word *Ohio* comes from the Iroquois word *Ohi-yo*, which translates to "beautiful river" or "great river."

Six current states—Ohio, Indiana, Illinois, Kentucky, Pennsylvania and West Virginia—made up the original Ohio Country, a vast land that formed the rich and fertile Ohio River Valley. The reason why the French and Indian War took place from 1756 until 1763 was that France claimed the Ohio River Valley as part of the New France colony. At the same time, the British considered it an extension of their Virginia colony. The valley area was rich in mineral resources and had a well-established fur-trading industry with the Native Americans. The Ohio River, which flowed into the Mississippi River, was regarded as an important transportation passageway for the territory and unfortunately was claimed by both the British and the French.

It should be noted that the French and Indian War was really a war between the French and British and included the thirteen British American colonies, France and the New France colony and at least ten Native American nations and confederacies, including the Iroquois, Cherokee, Wyandot, Shawnee and

The French and Indian War was fought between France and England over the rich Ohio Country. Painting by Frederick Coffay Yohn, circa 1905. *Cincinnati & Hamilton County Library.*

Algonquin. Also, as a result of the conflict over the Ohio Country, hostilities quickly escalated into what many historians regard as the first world war. It is better known as the Seven Years' War and involved France, Britain, Prussia, Brazil, Portugal, a number of German states, Russia, Spain, Sweden, Bengal Sultanate and the Mughal Empire.

Hostilities ended with the signing of the Treaty of Paris in 1763, which resulted in Great Britain emerging from the ordeal as the European holder of the Ohio Country. The Native Americans received no recognition in the Paris settlement, even though they were an important component of the economy. Since many Native Americans refused to accept British control, conflicts continued in the region for several decades. The Ohio Country was highly coveted and regarded as a jewel in the crown of America's economic future and expansion.

With the 1783 American Revolutionary War victory, the newly independent white Americans could now move west of the Appalachian Mountains. The American government's first order of business was to raise capital. Seeking land and opportunity, many people wanted to move westward. So, the Continental Congress and later U.S. federal government sold large tracts of land in the new Northwest Territory to real estate companies and individuals in what would later become the State of Ohio. Unfortunately, little regard was given to the fact that much of this land was still occupied by peaceful Native American tribes.

In 1787, Congress passed the Northwest Ordinance. In order to compensate Revolutionary War veterans and widows for their service and sacrifice, Congress agreed to grant portions of land in the Ohio wilderness. The amount of land varied based on previous military rank. For example, noncommissioned officers and privates received 100 acres, a colonel 500 acres, a brigadier 850 acres and a major general a healthy 1,100 acres for their service during the rebellion.

The conditions of the Northwest Ordinance ensured the new territory would be settled systematically. To begin with, slavery was forbidden, and every county had to be surveyed and laid out in thirty-six-square-mile townships. Also, one 640-acre section of each township was designated specially to fund public schools for education.

The first major attempt at settling the Ohio wilderness was orchestrated by the Ohio Company of Associates, formed in Boston in 1786. It was a speculative venture led by former army chaplain Manasseh Cutler and former generals Rufus Putnam and Benjamin Tupper. The Ohio Company pioneered the Northwest Territory's first permanent settlement, located on the banks of the Ohio River at Marietta. The company and its investors possessed 1.5 million acres of land, acquired from the Continental Congress with two $500,000 payments.

The business plan was simple and straightforward. First, a person crossed the treacherous Allegheny Mountains. Then, they cleared trees and stumps from the land purchased for one dollar per acre and constructed a home, which usually started with a log cabin. They then finally planted a spring crop to sustain their family. They would also have to survive an occasional altercation with the Shawnee, Cherokee and Delaware, who had been belligerently expelled from their ancestral lands.

Once the Ohio land settlers began producing extra produce and livestock that could be sold at market, a new problem presented itself. How do you efficiently transport goods to various markets for sale? Many early roads were just an expansion of previous Native American and deer trails. This proved extremely challenging, especially if your intent was to transport a large quantity of goods.

Colonel Ebenezer Zane (1747–1811) directed construction of a Northwest Territory frontier road, which came to be known as Zane's Trace. Constructed between 1796 and 1797, the 230-mile stretch of road started at

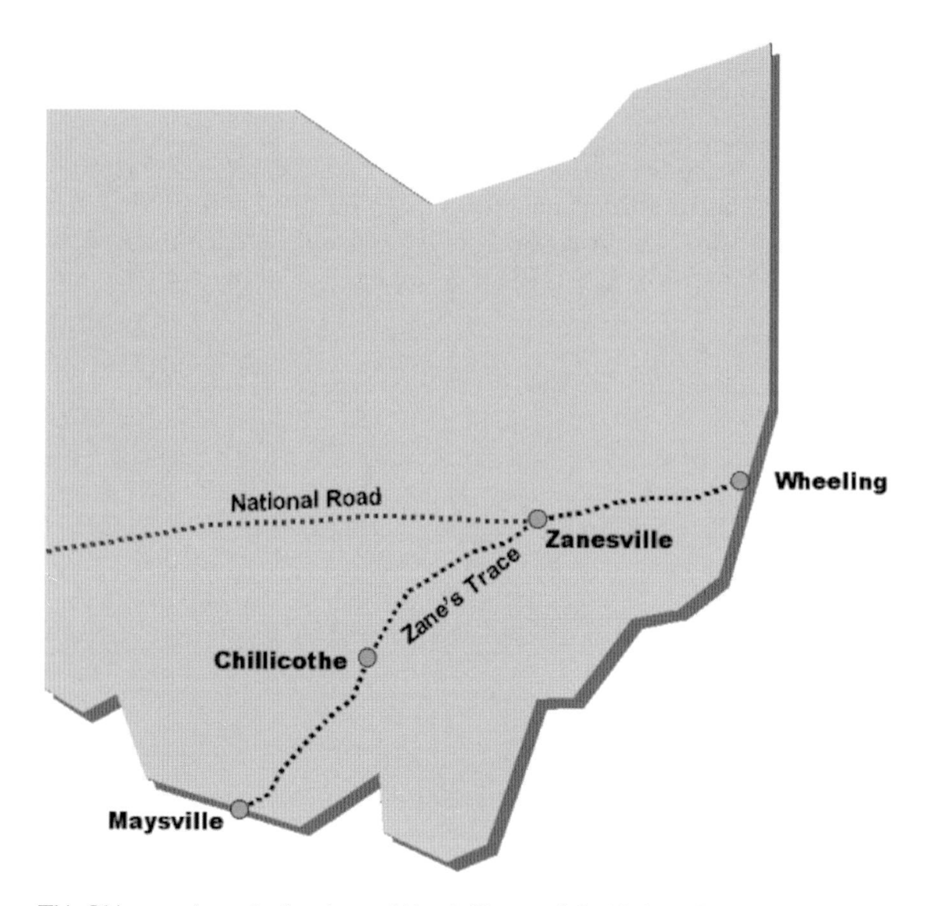

This Ohio map shows the frontier road Zane's Trace and the National Road, which later became US Highway 40. *Columbus Metropolitan Library.*

what is now Wheeling, West Virginia, and terminated at Maysville, Kentucky, running diagonally through the southeastern portion of Ohio. In 1796, Colonel Zane traveled to Washington to petition Congress for funding to finance the construction of a road to encourage settlement in the Ohio land portion of the Northwest Territory. He was looking to increase commerce while decreasing travel times to the Ohio River. It was clearly in the national interest for such a road to be constructed. Congress approved funding for the project in May 1796. The trace was constructed through heavily forested, hilly terrain and initially was not suited to transporting goods by wagon.

Ohio received statehood in 1803, and the state legislature levied a transportation tax in 1804 to upgrade and improve the entire route of the trace. Contracts were issued to widen the roadway and remove boulders and tree stumps. Between 1825 and 1830, the segment of Zane's Trace between Wheeling and Zanesville was rebuilt as part of the new and improved National Road.

In addition, in order to expedite the speed of commercial commerce and growth, the Ohio and Erie Canal was built. Construction began in 1825 and ended in 1832. It gave access to Cleveland and Lake Erie by way of the Cuyahoga River flowing to Akron and proceeding down to Columbus, Chillicothe and terminating at the Ohio River near Portsmouth, Ohio. Now Ohio had the means to prosper and grow a strong agricultural and industrial economy.

As the first state carved out of the Northwest Territory, Ohio was primed to become a leader in agriculture, manufacturing and applied technology and to emerge as an international industrial powerhouse.

2

TRANSPORTATION TO BUILD A BETTER ECONOMY

The nineteenth century saw Ohio blessed with an abundance of coal and oil for energy. It borders a Great Lake and had canals, an interstate river and a railroad system that could transport large quantities of raw materials to industrial centers such as Cleveland, Toledo, Akron/Canton, Dayton, Columbus and Cincinnati. All of this, coupled with curious problem-solving individuals with an aptitude for inventiveness, would catapult Ohio and the country into a leadership role in industrial development.

The natural geography of Ohio gave it an advantage that initially was used to establish an economy in new settlements such as Marietta and Cincinnati, which were strategically located on the Ohio River. The river gave these settlements access to the Mississippi River, which opened to the Gulf of Mexico and on to the Atlantic. This presented access to world markets for trading.

There is some evidence that Marietta, Ohio, was the point of origin for the frontier Midwest shipbuilding industry. In 1800, Marietta had easy access to excellent, inexpensive oak lumber from the untouched forests of the Ohio River Valley. Iron, coal and coal-tar pitch from southern Ohio were readily available, and mines in nearby Michigan provided copper. Westward migration and economic opportunity brought settlers with shipbuilding skills from the eastern coastal states to the Ohio Valley. A small operator could favorably navigate the Ohio and Mississippi Rivers by keelboat or flatboat carrying a small cargo. The varying seasonal river levels naturally placed limits on the size and type of ships that could feasibly

Early shipbuilding on the Ohio River in Marietta, Ohio. Shown here is a two-masted brig with square sails, circa 1795. *Washington County Public Library.*

be constructed at Marietta. The most popular design was the two-masted brig with square sails.

Marietta's shipbuilding industry on the Ohio River prospered until late 1807. The political juggling and conflicts that turned into the War of 1812 resulted in the Thomas Jefferson administration implementing the Embargo Act, which prohibiting trade with the British. While the embargo had little effect on the British, it destroyed the developing shipbuilding industry on the Ohio River. It took until the 1840s for the Ohio River shipbuilding industry to experience a rebirth.

Traveling by land wasn't easy or comfortable, but there was a wagon that was practical. Mennonite Germans produced the first Conestoga wagons in Pennsylvania around 1750. Following the Revolutionary War, this wagon was widely used for western migration to the opened Ohio area of the Northwest Territory and beyond. It had the ability to transport a maximum of eight tons of cargo and was pulled by oxen or a special, now-extinct breed of draft horses called Conestoga. The Conestoga wagon was a marvel in both design and construction and is regarded as the forerunner of freight and panel trucks that appeared at the end of the nineteenth century. The maintenance and servicing of these wagons were often done by livery stable owners. Some of these stables later developed into service garages and producers of parts and components that were supplied to many of Ohio's early auto manufacturers.

The portion of the National Road that passes through central Ohio is part of the first highway constructed entirely with federal financing. The road was authorized by the U.S. Congress in 1806 during the Jefferson administration, and construction began in Cumberland, Maryland, in 1811. By 1818, the new road had been completed to the Ohio River at Wheeling. From there, the road passes through Cambridge, Zanesville,

A Conestoga wagon, circa 1830. This was a horse-drawn freight wagon that originated in the eighteenth century and was designed for hauling freight over bad roads. *Museum of American History*.

The National Road passes through central Ohio and was the first federally funded highway in the nation. It started in Cumberland, Maryland, in 1811. *Library of Congress.*

Columbus and Springfield, Ohio, and continues to Richmond, Indiana, and on to Vandalia, Illinois.

The opening of the National Road prompted thousands of new settlers to head west over the Allegheny Mountains to settle the rich and fertile Ohio River Valley. Small communities along the National Road started to grow and prosper with the increase in population and commerce. Various stagecoach lines were used for commercial passenger travel on the National Road. A Conestoga wagon being pulled by a team of six draft horses could average approximately fifteen miles a day. Oxen were used for shorter trips. Travel by stagecoach could cover an average of sixty to seventy miles in one day.

As president and an experienced land surveyor, George Washington put forth the idea of building a canal that would run a route north from the Ohio River to Lake Erie. But not until 1822 did the Ohio legislature proceed to establish the Ohio Canal Commission and retain civil engineer James Geddes (1763–1838) to lay out the most efficient canal routes. Geddes came up with not one but two proposed routes for Ohio. In addition to a canal from the Cleveland Lake Erie area to Portsmouth on the Ohio River, he proposed a second canal, to be referred to as the Miami and Erie Canal. This canal began at Cincinnati by the Ohio River and went up the Great Miami River to Dayton. It was later extended farther north to the Maumee River just outside of Toledo, Ohio.

The initial financing for both projects came from loans by East Coast businessmen and investment bankers. The original cost estimate for the projects was $2.5 million but at completion came to a record $42 million. However, it should be noted that $25 million of the final cost accounted for interest paid by the state.

Construction of the canals required backbreaking labor. At one point, over 4,100 laborers, mostly poor Irish and German immigrants, were employed. The average wage was $7.50 a month plus food and a daily whiskey ration. Typhoid, malaria and dysentery were major problems and often resulted in death. The cost of the canal in the 1800s was often measured by how many men died per mile.

Work on the Ohio and Erie Canal began on July 4, 1825, near Newark, Ohio. Design specifications required the canals to span a minimum width of forty feet at water level and a depth of at least four feet and have a ten-foot-wide towpath. After the canal trenches were excavated, the walls and base bottoms were lined with Ohio sandstone or limestone. Southern Ohio canals remained open the entire year. Annually, extensive repairs would be required as a result of the seasonal flooding that damaged the locks, walls and towpaths. To prevent damage to the infrastructure due to the freeze and thaw cycle in northern Ohio, the canals were drained down to a foot of water in late November. Maintenance and repaired then took place in the spring. The Ohio and Erie Canal was completed in 1832.

The Miami and Erie Canal would take another twelve years to complete once the Ohio legislature allocated funding for the extension to Defiance and Lake Erie in 1830. On completion, the canals had various portions running through thirty-three of Ohio's eighty-eight counties. Eventually, the Ohio and Erie and the Miami and Erie Canal system consisted of 584 miles. The system gave Ohio a big advantage for growth and prosperity. Suddenly, the cost to ship goods from the developed East Coast to Ohio and vice versa declined substantially, from $125.00 per ton of freight to an average of $25.00 per ton. The travel time from Cleveland to Portsmouth along the Ohio and Erie Canal took just eighty hours. Traveling that same distance by stagecoach or horseback was quicker but more expensive. The cost to travel the same distance on a canal boat averaged $1.70 per person in the 1830s, which is equivalent to $56.00 in 2023.

Canal boats were maintained by their owners and averaged seventy-eight feet long and about fourteen feet wide. Mules were used primarily to haul freight boats, while two horses towed packets. It was usual for twelve-year-old boys to be hired as towpath walkers to lead the animals on the towpath.

A woman leading a set of mules pulling a canal boat along the Ohio Erie Canal towpath near Akron, circa 1880. *Akron-Summit County Library.*

They usually received a salary of twenty dollars per month. Towpath walkers were also charged with inspecting locks, aqueducts, dry docks and towpaths for any signs of wear. They watched for breaks in the canal to see that they were repaired quickly.

The canal system created exponential marketing opportunities for Ohio manufacturers, mills and farmers and made it possible to send goods and produce to New York City via Lake Erie, the Erie Canal and the Hudson River or to New Orleans via the Ohio and Mississippi Rivers. By 1860, Ohio led the nation in agricultural output with lumber, hogs, corn, wheat and coal being major products. Most of Ohio's canals remained in operation until the late 1800s. But by the 1850s, canals were facing serious competition from the railroad industry. Even though their service cost more, the railroads had several advantages over the canals. Railroad travel was by far much faster than canal service, and trains were not limited by the need for a water route. Because of these advantages, railroads quickly outpaced the canals and accelerated their demise.

3

OHIO INGENUITY FOR WORKING SMARTER AND HARDER

In 1900, there were plenty of capable, intelligent people in America working every day to build this country through their labor, trade or profession. But there is something unique about how Ohioans saw themselves in the role of helping to increase the standard of living for the general public and themselves. Long before American industrialist Henry John Kaiser (1882–1967) coined the slogan "Find the need and fill it," Ohioans were driven to do exactly that. A January 1899 edition of *Punch* magazine had an article about the coming twentieth century. A brief dialogue in a patent office between two individuals started with the question, "Isn't there a clerk who can examine patents?" The reply was straight forward, "Quite unnecessary, Sir. Everything that can be invented has been invented." Unfortunately for Charles H. Duell, the commissioner of the U.S. Patent Office in 1899, this statement has been incorrectly attributed to him and his legacy. But the fact of the matter is that Duell marveled with great expectations about the explosion of new patented inventions that would shape the twentieth century. He even stated that he only wished he could live his life over so he could witness the introduction of these advances.

The young men of the Civil War experienced a world of new gadgets, weapons and methods of transportation. Among the serious things suddenly in demand for the war effort were the ability to extract large quantities of raw materials and the ability to turn them into mass-produced products and goods. This was a different war, due to the need

A group of people pose for a photograph on a Baltimore Ohio Railroad steam engine, circa1858. *Geneva/Ashtabula County Library*.

for and ability to transport supplies and military ordnance using something more efficient than horses and wagons. The railroad was a key factor for both sides of the conflict, but it was used more effectively by the Union. Enter the Baltimore and Ohio Railroad (B&O), the first chartered railroad line in the United States and the backbone for transportation in the early 1800s. The line's construction began on July 4, 1828. From 1828 to 1861, the B&O expanded into thirteen states.

In addition to railroads, steam-powered ships and boats were critical in transporting troops and supplies, especially on the Ohio, Mississippi, Potomac, Cumberland and Tennessee Rivers and the Great Lakes. Ohio was in a perfect strategic location to take full advantage of the steam-powered transportation infrastructure already in place.

Wind and water power had been in use for centuries. But with the introduction of steam-driven technology, the idea of working smarter, not harder, was firmly on the minds of prospective entrepreneurs and industrialists. The truth is that more people in the non-labor-union sector of late nineteenth-century America could now work just as hard to produce more and to realize a larger profit for a few. But there was also the idea that the more one produced, the lower the final cost to consumers.

At the conclusion of a civil war that saw so many deaths due to unsanitary conditions, many individuals began thinking about improving the general quality of life through innovative tinkering and invention. Tinkering and repair of agricultural equipment was a normal way of farm life. But this was also a time when the idea of interchangeable components was being adopted as a best practice in American manufacturing, and this greatly contributed to transforming Ohio into an industrial giant in the post–Civil War years. But following the war, many people began looking deeper into the disciplines of science, engineering and business.

For many early Ohio inventors, their education had been very basic and centered on reading, writing and arithmetic. Ohio college professor William Holmes McGuffey's (1800–1873) series of elementary-school reading books, known as the McGuffey Readers, were used as a main primer for teaching reading to many nineteenth- and early twentieth-century individuals in America. McGuffey took a keen interest in public education during his time as a faculty member with Miami University in Oxford, Ohio. In 1835, he and the Cincinnati publishing house Truman and Smith entered into a contract to produce a primer textbook that covered a variety of basic subjects, a spelling book and four school readers. The first and second series were published the following year. The readers comprised didactic stories and excerpts from great books, because McGuffey felt that a proper elementary education required that a student be introduced to a wide range of topics to boost practical thinking. The readers were the standard texts in most states, with sales of over 120 million copies by 1920. Once a person had the rudimentary elements of a basic education, they were able to expand their knowledge by observation and experimentation. Most early Ohio inventors were extremely curious people, always asking questions, and they were avid readers of books and articles about facts, not fiction. These individuals were tactile learners who gained knowledge by experience and by doing hands-on tasks and generally being self-motivated learners.

ECLECTIC EDUCATIONAL SERIES.

McGUFFEY'S

NEW

FIRST ECLECTIC READER:

FOR YOUNG LEARNERS.

By WM. H. McGUFFEY, LL. D.

NEW-YORK ∴ CINCINNATI ∴ CHICAGO

AMERICAN BOOK COMPANY

FROM THE PRESS OF

VAN ANTWERP, BRAGG, & CO.

The McGuffey Reader was used as a primer textbook in America from the 1830s to the 1920s. The photo is circa 1835. *Cincinnati & Hamilton County Library.*

Post–Civil War manufacturing took shape in various ways. There are three main methods of manufacturing: job production, batch production and mass production (sometimes referred to as flow production). The type of manufacturing depended on the product being made and the type of raw material required. First, there is job production, which involves custom work on one product or a small production run made by a designated crew for a specified client or customer.

Batch production is used for products that are made in specified amounts, such as paint, and it can involve products grouped in different packaging. The cycle time—the time between each new batch and quality control—is very critical.

The third manufacturing method is mass production: rapid, high-volume duplication of a product using assembly-line technology to deliver partially completed components to a designated worker, who works on a single step. This is in lieu of having a skilled crew work to complete an entire product from beginning to end. Mass production includes fluid materials, which range from soft drinks to refined petroleum. These are products that are conveyed using pipes under pressure to transfer partially completed products between storage tanks and vessels. Bulk materials such as sand, iron ore, coal and agricultural grains are handled by chain, belt, pneumatic or auger conveyors.

Much of the technology was introduced when manufacturers switched and upgraded to electricity. Electricity gradually started being adopted as a power source for factories in the 1890s, after inventor Frank J. Sprague (1857–1934), introduced a practical DC motor. This practice was accelerated with the development of the AC motor by Nikola Tesla (1856–1943) and a host of other inventors. Between 1900 and 1929, there was a major push in switching from steam to electric power for factories, especially with the addition of central and regional generator stations that made electricity more cost-effective just prior to World War I. Electric

The Sprague Electric DC motor, seen here circa 1890, was widely used in late eighteenth-century factories, which were moving away from relying on steam power. *Library of Congress.*

motors proved to be exponentially more efficient than steam engine for power, because a central electric utility station generated more energy than a small factory steam engine.

The biggest advantage of early mass production was the ability to manufacture items that consumers regularly used. An example is the Ball Brothers Glass Company, founded by five brothers originally from Ohio. As a result of the natural-gas boom, they manufactured mason jars in Muncie, Indiana, not too far from the Ohio state line. At the turn of the last century, they decided to electrify their facility. With the introduction of automated electric glass-blowing machines, it was no longer necessary to employ the more than two hundred skilled craft glassblowers. Furthermore, an electric truck was introduced to carry 145 dozen glass jars at one time, replacing a hand truck with the capacity of just 6 dozen. Time-saving electric mixers were also put in service and replaced men who used to manually shovel sand and other raw materials into the gas-fired glass furnaces. The three dozen laborers used to move pallets and other loads were no long required after electric overhead cranes were installed.

Ohio developed itself as a comprehensive microeconomy in America that was ready to meet the consumer demands of a public looking to increase its living standard. The state was located in the heart of it all, and it was poised to have all that was needed to play a major role in transforming a vast wilderness into a modern, competitive society on the international stage.

4

OHIO'S INGENIOUS PROBLEM SOLVERS

Charles Hall

A native of Thompson, Ohio, Charles Martin Hall (1863–1914) patented an inexpensive method of producing aluminum on April 2, 1889. Prior to Hall's discovery, pure aluminum was rare and regarded as a precious metal. Aluminum is an abundant metal element, but it is found in a compound such as alum, instead of an ore that is easy to refine. Charles Hall graduated from Oberlin College in Oberlin, Ohio, in 1885 with a bachelor's degree in chemistry. He discovered a cost-effective way to produce pure aluminum, which opened up the metal being widely used commercially.

An older sister who was also a chemist, Julia Brainerd Hall (1859–1925), helped Charles in his research. The extent to which Julia was involved in her brother's research and the discovery of the Hall process has been disputed. But given that women in 1880s America were, foolishly, not regarded as able to think scientifically, these disputes are a product of the times—and in some cases even today. This is because most of the evidence has been destroyed or not taken seriously. Julia assisted her younger brother in fabricating many of his apparatuses and preparing the solutions and chemicals for his experiments. She also kept written records and notes of their work and prepared the required documentation for the patent application.

The chemist Charles Martin Hall made aluminum, which had been regarded as a precious metal, affordable. Hall is seen here about 1885. *Oberlin College Courtright Memorial Library.*

Hall's method of processing the aluminum was to pass an electric current through a nonmetallic conductor to separate the very conductive aluminum. A molten sodium fluoride compound was used as the nonmetallic conductor. At the same time in France, an inventor named Paul L.T. Heroult independently arrived at the same process. Fortunately, Hall had ample documentary evidence proving the date of his discovery, so the U.S. patent was awarded to him instead of to Heroult. But today, the use of electrolysis to extract aluminum is commonly referred to as the Hall-Heroult process.

In 1888, with the backing of financier and metallurgist Alfred E. Hunt, Charles Hall established the Pittsburgh Reduction Company with five other partners. The name was later changed to the Aluminum Company of America (ALCOA), and Hall served as the company vice-president. In 1859, aluminum was approximately $18 per pound. But by 1914, the year of his death, Charles Martin Hall had reduced the cost to $0.18 per pound, and it ceased being regarded as a precious metal. His discovery made him a wealthy man.

Hall served on the board of trustees of Oberlin College and, on his death, left an endowment of $5 million ($145 million in 2023).

Granville T. Woods

Granville Woods (1856–1910) was an African American inventor who dedicated his career to developing a wide range of inventions. Many of his inventions were essential to ensuring and increasing public safety related to railroad transportation throughout the United States and internationally.

Granville was born to Cyrus and Martha J. Woods in Columbus, Ohio, where he attended school until he was ten years old. At age thirteen, he began an apprenticeship in a machine shop and learned the machinist and blacksmith trades.

In 1872, Woods was hired as a fireman on the Danville and Southern Railroad and in time became a competent railroad engineer. In 1874, he relocated to Springfield, Illinois, and worked in the Springfield Iron Works rolling mill. During this time, he received two years of college-level training in electrical and mechanical engineering.

Alexander Graham Bell's company, American Bell Telephone Company, purchased the rights to one of Woods's patents, for a device that combined a telephone and a telegraph. This device allowed a telegraph station to send voice and telegraph messages through a single line. The proceeds Woods received from the sale of the rights allowed him to establish himself as a full-time inventor.

Relocating to Cincinnati in 1880, he set up his own business as an electrical engineer and inventor to invent, manufacture and market electrical apparatuses to improve railroad technology and associated equipment. Having a keen knowledge of thermal power and steam-driven engines, Woods's first patent was for an improved steam boiler furnace in 1889. Most of his patents, though, were for electrical devices. He developed over a dozen inventions to make electric railway cars safer and more efficient, as well as several devices that controlled the flow of electricity.

Wood's most noted work was the 1887 invention of the synchronous multiplex railway telegraph, which allowed communications between rail switch stations and moving trains. This amazing system allowed an engineer to know how close his train was to trains on the same track and helped reduce collisions. The technology involved was a unique variation of an induction telegraph that relied on ambient static electricity from existing telegraph cables to carry messages between railroad switch stations and moving trains. As a consequence of this work, collisions were reduced and many lives were saved. After receiving the multiplex telegraph patent, he restructured his company and renamed it the Woods Electric Company.

Granville T. Woods was a brilliant African American inventor whose products greatly increased public safety, especially as it relates to the railroad transportation industry. He is seen here about 1895. *Dayton and Montgomery County Library*.

Woods's success with the synchronous multiplex telegraph resulted in Thomas Edison launching two lawsuits against Woods on the grounds that he, not Woods, was the inventor of the multiplex telegraph. Woods won the court battle not once but twice. Following Edison's second defeat, he extended an offer to Woods: a high-paying, prominent position with the

The Synchronous Multiplex Railway Telegraph was patented by Woods in 1887 and served to reduce train collisions. *Dayton and Montgomery County Library.*

Edison Electric Light Company. Woods declined, because he didn't want to be controlled by Edison or under his thumb. Maintaining his independence was important.

He also invented a system for overhead electric conducting lines for railroads and streetcars. This was implemented in the development of overhead streetcar systems for growing urban cities such New York, Chicago and St. Louis.

In 1892, he moved his research operations to New York City, where he developed a safety dimmer system in 1896 for controlling electrical lights in opera houses and theaters. The system was economical, as it consumed 40 percent less electricity. In 1904, Woods patented an automatic air brake system for railroads that proved to be a significant improvement on the Westinghouse Air Brake.

At the age of fifty-six, Woods suffered a stroke on January 28, 1910, and died at Harlem Hospital two days later. Granville T. Woods's life's work made life easier and safer for people throughout the world, especially as it relates to railroad transportation. At the time of his death, he was an admired and highly respected inventor. He sold the rights to many of his inventions to large corporations such as Westinghouse and General Electric. Unfortunately, much of his financial gain went toward legal fees and defending the rights to his inventions in court.

THOMAS EDISON

Thomas Alva Edison (1847–1931) was a kid from Milan, Ohio, who grew up to become a highly accomplished inventor and industrial entrepreneur. Historically, it is important to acknowledge that Edison's formal education amounted to just three months and, subsequently, being homeschooled by his mother. But if there was a college degree in curiosity, he would have doctorates with honors, along with degrees in electrical engineering, industrial technology, material science, education, biology, chemistry, physics and business administration.

Thomas A. Edison as an Ohio inventor had the greatest impact on establishing the modern twentieth-century industrialized world of technology and entertainment. He is seen here circa 1915. *Huron County Community Library.*

Although Edison did most of his inventive work in Menlo Park and West Orange, New Jersey, and Fort Myers, Florida, it was his early childhood experience and nurturing in Ohio that launched him on a path to having a worldwide impact on the modern twentieth-century industrialized world.

His general approach to inventing, innovating and problem-solving was to apply science in a collaboration with the talents of other technicians, researchers, tradesmen and tinkerers like himself. During his life as an inventor, he advanced the science of electricity by perfecting machines for sound recording, motion pictures, power generation/utility and mass communication. His work led to the invention of the early electric light bulb, the phonograph and the motion picture camera. Edison had some 1,093 patents in his name or jointly shared with other inventors on his staff. He is regarded as the most prolific inventor of the nineteenth and twentieth centuries.

To understand the impact that Edison's work had on the general way of life that had been in place for thousands of years, consider the following example. Fireplaces and candles were the standard for lighting a home after dark, but generally, people went to bed at nightfall. Early Industrial-Age factory production hours were also limited due to a lack of light. Factories were now able to operate a full twenty-four hours around the clock after an electric light system was introduced. Municipal electric companies using Edison steam powered generators later provided power and light to homes. Much has been written about Edison, but regarding the subject of Ohio, innovation, invention and manufacturing, Edison will always get a standing ovation.

WRIGHT BROTHERS

Brothers Orville Wright (1871–1948) and Wilbur Wright (1867–1912) jointly invented, fabricated and successfully flew the world's first successful motor-operated airplane in 1903, just south of Kitty Hawk, North Carolina. In addition to owning a printing shop, the brothers started building bicycles in 1896. In time, the profits from the printing shop and the bicycle shop would fund the Wright brothers' aviation experiments from 1899 to 1905. Their experience in designing and fabricating bicycles proved ideal preparation to building airplanes.

They were familiar with the work of German glider pioneer Otto Lilienthal, and the news of his death in a glider crash in 1896 was the start of their serious interest in aviation. They introduce themselves to Octave Chanute, a civil engineer and authority on aeronautics, who would work with the brothers during the critical development years of 1900–5. The Wright brothers' plan for building an airplane was similar to Lilienthal's. But the Wrights saw that his methods of balance and control were seriously inadequate and required a different solution. Using aeronautical data on lift published by Lilienthal, the Wrights based their design on the Chanute-Herring biplane hang glider, which flew successfully during experiments in 1896.

By 1903, the brothers had completed the powered *Wright Flyer* using lightweight spruce wood and a strong muslin fabric as a surface covering. They also designed and fabricated the wooden propellers, which had a peak

Dayton bicycle manufacturers Wilbur Wright (*left*) and Orville Wright, seen here about 1905, invented "powered" flight and maneuverability. *Dayton and Montgomery County Library.*

efficiency of 82 percent. No engine manufacturers could meet the design criteria for a lightweight engine. So the Wrights worked with their own shop mechanic, Charlie Taylor, and, in just six weeks, built the engine.

The *Wright Flyer*'s official first flight took place on December 17, 1903, with Orville at the controls and Wilbur running alongside at the wing tip. Of the five people who witnessed the flight, John T. Daniels snapped the "first flight" photo using Orville's pre-positioned camera.

In 1904 and 1905, the brothers worked to improve their flyer and enable it to make more aerodynamic flights. This was the *Wright Flyer II*. This was followed by the *Wright Flyer III*, which was essentially the first practical fixed-wing aircraft. Another one of the Wright brothers' contributions to aviation was their innovative three-axis control system, which allowed the pilot to steer the aircraft more effectively while maintaining its equilibrium. Even today, this method remains standard on all fixed-wing aircraft. Wilbur and Orville always focused their attention on developing a reliable means of pilot control as the way to make flying safe.

After the Wright brothers built the *Wright Flyer* in Dayton, it was tested on December 17, 1903, near kitty Hawk, North Carolina. *Dayton and Montgomery County Library.*

The first 1903 *Flyer* cost $1,000 (equivalent to $32,000 in 2023). This is in contrast to the $50,000 ($1.8 million in 2023) in military funding that Samuel Langley received to develop his man-carrying Great Aerodrome flying machine. Nine days before the Kitty Hawk flight, Langley's Aerodrome never got airborne and plunged into the Potomac River.

After their first successful flight, the Wright brothers were still regarded as nobodies from Ohio and endured skepticism from the international aeronautical community, especially the French. Wilbur started giving public demonstrations on August 8, 1908, at a horse track outside Le Mans, France. His first flight lasted one minute, forty-five seconds, but his ability to make banking turns and fly in circles amazed the crowd. Over the next few days, Wilbur executed several challenging flights that included making a figure-eight and demonstrating his ability as a pilot and the capability of his flying machine. The demonstration surpassed all other pioneering aircraft and aviators of the day.Wilbur's demonstration near Le Mans thrilled the French public. The Wright brothers were an instant worldwide sensation, and apologies and praises were issued by former detractors.

Four years later, in 1912, Wilbur fell ill and died of typhoid fever. Orville died in 1948 but lived to see the dawn of the atomic age and jet-powered flight.

CHARLES KETTERING

Charles F. Kettering (1876–1958) was a classic late nineteenth-century curious Ohio boy who, because of his poor eyesight, was finally able to graduate from Ohio State University in 1904 at age twenty-seven with an electrical engineering degree. Kettering relocated to Dayton, where he was hired by a National Cash Register Company manager, Edward Deeds. Deeds assigned Kettering to the engineering research department and placed him in charge of developing the first electric cash register.

In 1909, Kettering and Deeds established the Dayton Engineering Laboratories Company, better known as Delco. At Delco, Kettering also perfected an electric lighting system for automobiles to allow for safe night driving. They started working after hours in a barn at Deeds's north Dayton home. Kettering, working with a group of men known as the "Barn Gang," developed the first electric ignition system and practical self-starter for an automobile. The device, first installed on a Cadillac in 1911, alleviated the need to turn a hand crank at the front of the car to start the engine.

The first car manufacturer to buy the Delco starters was William Durant's General Motors. Kettering's starter and lights became standard equipment on all GM vehicles. As a result, General Motors acquired Delco in 1916. Kettering was installed as general manager of the new research division

Charles Kettering working on the electric self-starter in his Deeds barn workshop, circa 1910. *Dayton and Montgomery County Library.*

of GM. Toward the end of World War I, Kettering's team developed the Kettering Bug, an experimental, unmanned aerial torpedo that is considered the forerunner of today's cruise missiles. In 1920, he was appointed as a vice-president of General Motors. Kettering and his staff of engineers and chemists continued to improve and develop automobile technology, including automatic transmissions, leaded gasoline, spark plugs and four-wheel brakes. GM's research division under his leadership also developed improved diesel engines, safety glass and the refrigerant known as Freon, which was used in GM's Frigidaire division. Kettering's south Dayton residence was the first house in America to be equipped with whole-house electric air-conditioning. Kettering, who was granted 140 patents, retired from General Motors in 1947.

Charles Franklin Kettering, seen here about 1925, was an inventor known for his ability to make existing inventions even better and more efficient. *Dayton and Montgomery County Library.*

Charles F. Kettering was an active and serious philanthropist. In 1945, he and Alfred Sloan, the president of General Motors, established the Sloan-Kettering Institute for Cancer Research in New York City. In 1980, Kettering was inducted into the National Inventors Hall of Fame, headquartered in North Canton, Ohio.

5

INNOVATIONS IN WASHBOARDS, PAPER, GLASS AND ARITHMETIC

The Columbus Washboard Company

The Columbus Washboard Company was established in 1895 by Frederic Martin Sr. His washboards were handmade and had a limited market. In 1925, Frederic Martin Jr. acquired the company from his father with assets that consisted of a few saws, a metal crimping machine and the 1907 patent and trademark for the Bear Easy washboard. Frederic Jr. and his wife, Margaret, were involved in the business of Columbus Washboard from the date of purchase in 1925 until their deaths in 1987 and 1988, respectively. From their start in 1925 until their deaths, Frederic and Margaret Martin manufactured and sold over twenty-three million washboards, including such trademarked brand names as Dubl Handi, Maid-Rite Silver, Maid-Rite Brass, Sunnyland and Crystal Cascade. Fortunately, most of the original dies and designs have been maintained by the company.

Today, the company is located in Logan, Ohio, and has recently relocated to a new permanent location at 4 East Main Street. The owners are Jacqui Barnett, Larry and Joyce Gerstner and James Martin, and the factory is managed by Diane Hopkins. Production is in the capable hands of Lisa Jarrell and Linda Blackburn. The washboards are still assembled by hand on the original presses used by the Martin family since 1895. Lisa and Linda also hand-pack all orders to a high standard for shipping.

Left: A standard family-size washboard as made by the Columbus Washboard Company, circa 1970. *Courtesy Columbus Washboard Company Collection.*

Below: A Columbus Washboard Company staff member manually assembling a washboard, circa 2010. *Courtesy Columbus Washboard Company Collection.*

The company maintains a loyal base of customers that includes the Amish community, which continues to use washboards for their laundry. Columbus Washboard Company is the last manufacturer of genuine usable washboards in America. The factory is open daily except Sunday for public tours and welcomes guests from 9:00 a.m. to 3:00 pm.

Mead Paper

In the late twentieth century, one of the world's largest manufacturers of paper was the Mead Corporation of Dayton, Ohio. Colonel Daniel Mead (1817–1891) and his partners established the Ellis, Chafflin & Company in Dayton to make books and manufactured paper in 1846. By 1881, Mead had bought out his last partner and established the Mead Paper Company in 1882. In 1890, he acquired the Ingham Mill in Chillicothe. At the time of Colonel Mead's death in 1891, his company was one of the largest paper manufacturers in the United States.

The company management was passed to sons Charles and Harry, who became, respectively, president and vice-president. Despite the fact that Daniel Mead had left a thriving business, his sons' misuse of company funds brought Mead Paper Company to a near total collapse in 1905. Banker-trustees turned to George Mead, Harry Mead's business-minded son, to take over leadership. George agreed to take on the task of restructuring the family business. He renamed the business Mead Pulp and Paper Company and was named vice-president and general manager.

In 1906, Mead Pulp and Paper made a public stock offering, and in June 1906, the original Dayton mill was shut down and all paper milling operations shifted to the Ingham Mill in Chillicothe. The 1907 recession, coupled with the cost of moving, almost destroyed Mead again, but the proceeds from selling the Dayton mill property saved the company.

Mead began expanding through acquisitions and in 1917 purchased the Peerless Paper Company of Dayton. George Mead wanted to limit the general product line, because in 1905, floor operations produced fifteen different grades of paper. In this way, profits would be maximized if each machine produced one type of paper, instead of retooling and changing production methods for different papers.

Mead secured a five-year contract in 1917 to produce magazine paper for Crowell Publishing Company. This alone required 75 percent of the

Woodchips being loaded into a bash digester at Mead Paper in Chillicothe, circa 1938. *Dayton and Montgomery County Library.*

Chillicothe mill's production. Crowell remained Mead's primary customer for the balance of the decade. In 1918, a separate firm, the Management Engineering and Development Company, was established in Dayton to oversee the plant engineering of new Mead.

In 1920, Mead fully acquired Kingsport Pulp Corporation in Kingsport, Tennessee. This plant began white paper production in 1923. By 1925, Mead researchers developed a semi-chemical pulping process that enabled wood chips in which tannin had been previously extracted to be recycled into paperboard. This allowed the paperboard business to be expanded in the late 1920s and the acquisition of mills throughout the Appalachian region that produced corrugating medium from wood waste. In 1927, the Mead Paperboard Corporation was established as a holding company for the paperboard operations, which included the Sylvia Paperboard Company, the Harriman Company, the Southern Extract Company and the Chillicothe Company.

On February 17, 1930, with George Mead as president, the Mead Corporation was incorporated. The corporation's subsidiary operations included the Mead Pulp and Paper Company, the Mead Paperboard Corporation and the Management Engineering and Development Company. The corporation had a total of one thousand employees and plants in four states.

In 1942, George Mead became chairman of the board and Sydney Fergusen was appointed the corporation's president. In the same year, Mead acquired a small white paper mill from the Escanaba Paper Company of Michigan. In time, the Escanaba mill became one of Mead's largest operations. Although Mead had continued production at an accelerated pace to meet domestic and overseas demand, wartime price and profit controls, in addition to raw material shortages, stunted the company's growth.

Immediately following World War II, Mead's brown paper division made plans to erect a kraft linerboard mill to replace an older facility. Mead proceeded to firmly entrench itself into paperboard manufacturing by forming a joint project with Inland Container Corporation. The two companies collaborated in 1946 to start the Macon Kraft Company to erect and operate a paperboard mill in Macon, Georgia. The acquisition of the Atlanta Paper Company that same year led Mead into the packaging business and was the forerunner of Mead's packaging division. This division patented the paper six-pack carrier for bottled beverages and beer and became the world's largest supplier of paperboard beverage packaging.

In 1960, Mead's accelerated expansion in paperboard manufacturing prompted the Federal Trade Commission (FTC) to initiate a complaint against Mead, alleging that the company's growth since 1956 was anticompetitive. A settlement was reached in 1965 with the FTC when Mead signed a consent decree agreeing to sell off seven of its facilities over a five-year period. The agreement also included a ten-year moratorium on paperboard acquisitions. Mead operated diverse operations that included furniture factories, foundries and coal mines throughout the 1970s.

A Mead mill fourdrinier machine is removing the water from the paper pulp, circa 1938. *Dayton and Montgomery County Library.*

During the early 1980s, profits began to slip from the 1979 peak. But business improved in 1984, when Mead's electronic information-retrieval services began to see a profit. Mead Data Central Inc. (MDC), a software subsidiary whose primary product was LEXIS (a service that made case law and statutes available through online computer searches), was growing by 43 percent a year. LEXIS was introduced in 1973, and by the end of the decade, it had nearly 75 percent of the computerized legal research market.

Mead's staff also restructured its paperboard operations to focus more on the production of coated board. In 1988, Mead took over complete control of the Phenix City, Alabama coated board mill. In 1991, Mead completed an expansion of this mill, which added 370,000 tons of coated board annually. Mead also sold its share of the Brunswick pulp and paper mill in August 1988 and sold its recycled products business to Rock-Tenn Company in 1988.

In 1992, Steven Mason became company president and unveiled a three-year improvement plan aimed at increasing both productivity and customer satisfaction. Part of the plan included dismissing nearly one thousand employees while setting up a special reserve fund for items such as severance pay, relocation, counseling and outplacement. By the end of 1996, Mason had successfully achieved an overall productivity gain of 12 percent since 1992.

Mason also refocused Mead on core value-added forest products. Mead sold its Kingsport, Tennessee uncoated paper mill in 1995, which had been losing money. But its largest divestment was the December 1994 sale of Mead Data Central (MDC) to Anglo-Dutch publishing giant Reed Elsevier. This sale took Mead completely out of the electronic publishing business. Mead could now focus on its three core areas of operation: coated paper; packaging and paperboard; and distribution and school/office supplies. The proceeds from the MDC sale were used to reduce debt and make stock repurchases.

Mead's 150th anniversary in 1996 was marked with solid revenues of $4.71 billion. The Mead Corporation became known as MeadWestvaco following a January 2002 merger between the Mead Corporation of Dayton, Ohio, and Westvaco. In 2006, the Dayton corporate office was relocated to Richmond, Virginia, until 2015, when MeadWestvaco and Georgia-based RockTenn merged to form WestRock.

LIBBEY GLASS COMPANY

Beginning in 1878, William L. Libbey (1826–1883), who had been a manager for the New England Glass Company of East Cambridge, Massachusetts, leased the facilities of his former employer and established his own glass-blowing factory. To help manage the company, Libbey enlisted the help of his son Edward Drummond Libbey (1854–1925), who was twenty-four years old and had previously planned to become a minister. Following the death of his father in 1883, Edward took over running his father's company and was quickly confronted with a variety of financial challenges. The worst was in 1886, when a strike was called by the American Flint Glass Workers Union. The grievance was for an increase in wages at a time when energy costs were reducing profit margins.

Meanwhile, the city of Toledo, Ohio, was looking to attract new businesses to the community to take advantage of the regional natural gas boom of the 1880s. Libbey was offered an incentive package that consisted of land for a new factory and an additional fifty lots for employee housing, as well as $100,000 ($3.5 million in 2023), to construct a new manufacturing plant. Edward Libbey proceeded to take full advantage of the opportunity to financially turn his father's business around. In August 1888, he traveled from Massachusetts by train with 250 skilled glassblowers to Toledo. Unfortunately, adapting to the midwestern way of life proved difficult for many of his employees, who chose to return to New England.

Edward Drummond Libbey, shown here about 1885, was the industrial pioneer who established the glass industry in Toledo. *University of Toledo Library*.

In hopes of resolving his employee problem, Libbey traveled to Wheeling, West Virginia, looking for new skilled workers at the Hobbs & Brockunier glass factory. There he met Michael J. Owens, who, since the age of ten, had been employed in a glass factory. Libbey saw in Owens the leadership skills he needed for a manager of his new factory and offered him the position, which Owens accepted. Owens was general manager of the Toledo facility and an additional glass factory in Findlay, Ohio, that manufactured bulbs for the new electric light industry.

Despite Libbey's vigilance, the company continued to have financial problems. Libbey decided to rename the business the Libbey Glass Company, with the expectation of making the company nationally recognized as a leader in the industry. The Libbey Glass Company manufactured bottles, containers and window glass and was best known for producing exquisite cut glass. Libbey wanted to invest $200,000 (equal to $7 million in 2023) of the company's money to construct an actual glass furnace at the 1893 World's Columbian Exposition in Chicago. He reasoned that such a grand advertising display would introduce the entire nation to his company. However, the board of directors didn't comprehend the full merits of Libbey's creative idea.

But Libbey would not be deterred and was able to raise private funding to build his world's fair exhibit. Libbey made Michael Owens the manager in charge of the Chicago exhibit, which initially didn't draw much of a crowd. So Libbey allowed the fairgoers to apply their admission fee to the purchase of glass baubles and trinkets inscribed with the Libbey name.

The Libbey Exhibition Pavilion at the 1893 World's Columbian Exposition in Chicago, where the name *Libbey* was introduced to the nation. *University of Toledo Library.*

Consequently, the exhibit began drawing huge crowds. One reason for the crowds was that many were coming to get a glimpse of a dress fashioned using spun glass fabric for the Broadway actress Georgia Cayvan (1857–1906). Spun glass fabric was eventually used to make products like thermal insulation and was the foundation for a major new glass industry pioneered by Libbey's company. The success of Libbey's exhibit at the Columbian Exposition turned the company into a profitable enterprise. At the 1904 St. Louis World's Fair, Libbey exhibited brilliant cut glass pieces and featured an astounding punch bowl that was regarded at that time as the largest piece of cut glass in the world.

Libbey's wife, Florence Scott, was the granddaughter of one of the University of Toledo's founders, Jesup Scott. Florence had a great appreciation for fine art and, in 1901, along with her husband, established the Toledo Museum of Art. In 1925, at the age of seventy-one, Edward D. Libbey died of pneumonia. Most of his estate was left to establish an endowment for the internationally recognized art museum, and this gift funds the museum's operations to this day. His company, the Libbey Glass Company, is regarded as the foundation of all glass-related companies currently in Toledo.

OWENS-ILLINOIS

Industrialist and inventor Michael Owens, shown here about 1920, knew the glass industry from the ground up, and he revolutionized glass production. *University of Toledo Library.*

Owens-Illinois of Toledo. Ohio, is one of the leading manufacturers of packaging goods internationally and the largest manufacturer of glass containers in North America, South America and India. It is the second largest in Europe. It is estimated that half of all glass containers made worldwide are manufactured by Owens-Illinois, its affiliates or its licensees. The company's plastics subsidiaries produce a wide range of plastic packaging, including containers, closures, trigger sprayers, finger pumps, pharmaceutical prescription containers and multipack carriers for canned beverages.

The Owens-Illinois Glass Company was founded by Michael Owens (1859–1923), who played a major role in the glass manufacturing industry when he invented the automatic glass-blowing machine. A native of Mason County, West Virginia, Owens started a glassware apprenticeship at the age of ten with the glass firm Hobbs & Brockunier and Company in Wheeling. By age fifteen, he was a master glassblower. He was later recruited by glass manufacturing industrialist Edward Drummond Libbey to work for his Toledo-based factory. Libbey appointed Owens as a manager responsible for engineering and assembling machines that would automate the production line.

In 1890, workers at the Corning Glass Works in Upstate New York went on strike. The immediate result was a serious shortage of light bulbs for the Edison General Electric Company for the expanding electrification of American communities. Libbey was approached to take over light-bulb production for the duration of the strike. In Findlay, Ohio, Libbey leased a closed glass plant to produce the bulbs and appointed Owens as plant superintendent. For one year and five months, the plant produced light bulbs at a significant profit, wiping out the debt of the struggling Toledo Libbey Glass plant and allowing it to continue operating. Owens and general manager Sol Richardson were credited with saving Libbey's company.

Another success for Owens was the 1893 Columbian Exposition exhibit, where he oversaw operations of the Libbey glass factory. That same year,

Owens began work on developing a machine to automate light-bulb production. He set up an engineering laboratory in Libbey Glass and focused on developing a semiautomatic machine that could blow bulbs into molds. The machine was designed with five rotating arms with a device similar to that of a blowpipe with a mold at the end. The pipe would pick up a measure of molten glass, and compressed air would blow the glass into the mold. In one hour, the machine could produce four hundred bulbs. Although more labor was required to produce bulbs using this method, the labor did not have to be skilled, and this minimized costs.

In 1895, Libbey and Owens jointly received a patent for this semiautomatic machine and started a new company to market Owens's innovations. Owens went to work to modify the bulb-making machine into one that could also make glass tumblers and glass lamp chimneys. In 1898, Owens focused his attention on developing a fully automatic machine to manufacture glass bottles. Over a period of five years, the automatic bottle machine was successfully developed with an investment of over $500,000 ($17 million in 2023).

In 1904, Owens was granted a patent for an automatic machine to manufacture glass bottles. His invention completely automated the entire production process, from measuring the required amount of molten glass to blowing the glass into its final bottle shape. The machine produced bottles at a rate of 240 per minute while decreasing the labor required by 70 to 80 percent.

In 1903, just prior to receiving a patent for an automatic bottle machine, Owens, with financial help from Libbey, established the Owens Bottle Machine Company in Toledo to license the new invention. This firm initially manufactured Owens's bottle machine. Owens's invention of a bottle-making machine not only earned him fame but also contributed to Toledo's importance in the glass industry. By 1919, the firm had begun to manufacture bottles, and the company changed its name to the Owens Bottle Company. Advanced models of the bottle machine were also engineered; over three hundred different models of the machine were manufactured between 1908 and 1927. The company grew quickly and acquired the Illinois Glass Company in 1929. The Owens Bottle Company became known as the Owens-Illinois Glass Company that year. Ten years later, Owens and Libbey's establishment merged with the Illinois Glass Company to become the Owens-Illinois Glass Company. In 1965, the company changed its name to simply Owens-Illinois Inc.

The Libbey Glass Company was acquired by Owens-Illinois in 1935 and was now in the business of consumer tableware. The Libbey division

The Owens automatic glass bottle-making machine was patented in 1904. *University of Toledo Library.*

continued to turn out dishes, pitchers and bowls. Meanwhile, the ceramic engineers of Owens-Illinois were working to develop products using glass fibers. On learning that its main competitor, Corning Glass, was pursuing similar research, the two companies agreed to a collaboration. In 1938, Owens-Corning Fiberglass was born, and it proceeded to develop marketable fiberglass products. Together, Owens and Corning had monopolized fiberglass technology and were realizing a substantial return. A 1949 antitrust ruling that barred Owens and Corning from having complete control of Owens-Corning. As a result, the joint venture went public in 1952, and one-third share was distributed each to Owens, Corning and the general public. Both Owens-Illinois and Corning Glass later sold their interest in Owens-Corning.

When World War II ended, Owens-Illinois functioned primarily as a glassmaker. Several antitrust rulings in the late 1940s restricted corporations like Owens-Illinois from increasing their market share by acquiring subsidiaries related to their own industry. That meant that any growth through acquisitions would have to be in fields other than the glass industry. In the late 1960s, the company acquired Lily Tulip Cups Company, which manufactured products such as wax-lined milk cartons and disposable cups. This type of diversification prompted Owens-Illinois Glass to retire the word *glass* from its official company name and to become simply Owens-Illinois Inc.

When general beverage sales began to level off in the 1970s, the container and related industries began feeling the impact of a full-scale worldwide recession. Many large breweries and soft-drink producers began manufacturing their own containers, while some can and bottle manufacturers decided to increase the size of their container-producing facilities. Now they were confronted with overcapacity and reduced prices. The problem was especially felt in bottle manufacturing, because of the more labor-intensive production. A wholesale modernization was needed. So the company moved to divest itself of a number of marginal interests and committed to invest $911 million ($3.2 billion in 2023) into a four-year plant modernization program.

In addition to investing heavily in research, production methods were developed that significantly reduced the labor required to turn out finished glass products. A total of forty-eight plants were closed and seventeen thousand employees were dismissed; the jobs of forty-six thousand employees were saved by streamlining general operations.

A number of Owens-Illinois' competitors failed to invest the money necessary to a competitive edge. Thatcher Glass, which produced milk bottles and was once number two in the industry, went bankrupt in 1985 after refusing to rebuild its old furnaces and install new technology. Other manufacturers, such as Anchor-Hocking and Glass Containers Corporation of Lancaster, Ohio, closed their doors due to similar circumstances.

Today, Owens-Illinois is known simply as O-I and is still a multinational and diverse firm. Its global headquarters is in Perrysburg, Ohio. As of 2022, it is made up of twenty-five plants in nine countries.

HEISEY GLASS COMPANY

Augustus H. Heisey (1842–1922) established a glass factory in Newark, Ohio, in 1896, during America's Gilded Age. As a young apprentice, Heisey learned the glass-blowing trade. He served in the Union army during the Civil War and afterward found himself working as a salesman for the glass manufacturer Ripley and Company. Heisey spent the next two decades learning the comprehensive details of the glass industry. Following a brief

Above: The Heisey Glass Company operated in Newark, Ohio, from 1896 to 1957. The factory is seen here about 1915. *Licking County Library*.

Right: An original Heisey Glass Company crystal clear glass pitcher, circa 1920. *Licking County Library*.

time out West working in the mining industry, Heisey devised a strategic plan to establish his own glass business. The Newark, Ohio Board of Trade was looking to recruit new industries to take advantage of the inexpensive labor and the abundant natural gas resources in the area. Accepting the board of trade's offer, Heisey decided to plant roots in Newark and began building a factory in 1895. In April 1896, the new factory was placed in service with one sixteen-pot furnace. Eventually, the factory was equipped with three furnaces and a staff of 690 full-time workers.

The Heisey Glass Company trademark design consisted of an *H* centered inside a diamond. This logo was used from 1900 until the official closing in 1957. By the end of the Victorian era, fine glassware was in great demand, and Heisey would market his creations internationally.

As early as 1910, the company pioneered advertising glassware in national magazines. Heisey began manufacturing etched-glass designs in 1914 and started making blown ware, which was known as Heisey's American Crystal. Going beyond turning out the traditional pulled stemware, Heisey was the first glass company to manufacture fancy pressed stems. The results of using this process proved a big hit with the public. Today, most hand-wrought stemware is manufactured using this procedure.

After the elder Heisey's death, his son Wilson, a trained chemist, took over management of the company. Wilson contributed a wide range of new colors to the company's glass line, including pastel colors that proved very popular with the public in the 1920s and early 1930s. Colored glass faded from the market following Wilson Heisey's passing in 1942. World War II put a strain on the glass industry. But following the war and under Clarence Heisey's management, the company began manufacturing its famous line of figurines. In the 1950s, foreign competition was starting to have an adverse impact on the American glass industry. Unfortunately, Heisey Glass Company had to discontinue all operations and closed its doors in 1957. The following year, the Imperial Glass Corporation of Bellaire, Ohio, acquired Heisey's remaining assets. Using the Heisey glass molds, Imperial Glass manufactured minimal amounts of glassware until August 1984, when the company went bankrupt.

NATIONAL CASH REGISTER COMPANY

The history of the National Cash Register Company began with Dayton, Ohio saloonkeeper James Jacob Ritty (1836–1918). Ritty was getting fed up with how several dishonest staff members would pocket money designated to pay for the food, drink and cigars. In 1878, while on a steamship, Ritty observed a device that counted the rotation of the ship's propeller. He started thinking that a similar device could have merit if applied to recording the cash transactions at his saloon.

On returning to Dayton, Ritty using the mechanical skills of his brother John and went to work creating a cash-counting device. The prototype didn't have a cash drawer but worked by pushing a key representing a specific amount of money. James and John Ritty call their invention Ritty's Incorruptible Cashier and were issued a patent on November 4, 1879.

The Rittys tried to manufacture and market their cash-registering device, with little success. In 1881, they sold their full interest in the cash-register business to a group of investors that included John and Frank Patterson.

John H. Patterson, owner and founder of the National Cash Register Company, shown here about 1912. *Dayton and Montgomery County Library.*

Initially, the firm was called the National Manufacturing Company. By 1884, John H. Patterson (1844–1922) assumed majority ownership of the company and renamed it the National Cash Register Company.

As a business pioneer, Patterson transformed National Cash Register into a modern American company by implementing aggressive marketing plans and business methodologies. In 1893, he opened the first sales staff training school. He also invented the formal sales training academy called Sugar Camp and implemented a comprehensive social welfare program for the entire company staff.

Initially, the company grew slowly, and it manufactured 16,000 cash registers during the first decade in business. But under Patterson's leadership, by 1914, the company was manufacturing over 110,000 cash registers annually and was employing over 6,500 workers.

Patterson was able to attract highly capable and very talented staff, including Thomas J. Watson Sr. (1874–1956) as the company's sales manager (later the chairman and CEO of IBM). He also hired engineer Edward Andrew Deeds (1874–1960), who served as president of the company after Patterson's death. In 1904, Deeds hired Charles F. Kettering (1876–1958), an innovative and capable electric engineer. In 1906, Kettering created the first electric motorized cash register. Deeds and Kettering later established the Dayton Engineering Laboratories Company, better known as DELCO.

National Cash Register was regarded as "America's model factory," and Patterson had an international reputation for looking after the general welfare of the staff in the workplace and beyond. He provided workers with benefits, including rest periods; hot meals in clean dining rooms; and medical care from doctors, nurses and volunteer caretakers, health education and a comprehensive infirmary. Machine operators sat on ergonomic chairs with back support instead of stools. Workers had indoor sanitary facilities with hot and cold running water. Patterson had indoor mechanical ventilation systems installed to provide cool, clean air to the shop workers.

Immediately after the great Dayton flood of 1913, all manufacturing ceased in order to focus attention on the company's Citizen Relief Committee. This group housed flood refugees; provided food, clean drinking water and

National Cash Register Company's modern Dayton, Ohio campus, circa 1920. *Wright State University.*

Cash registers being assembled in a building designed for maximum ergonomic efficiency, circa 1915. *Wright State University.*

NCR was a manufacturing complex designed as a "daylight factory" and built with floor-to-ceiling operable windows to allow fresh air ventilation. It is seen here circa 1910. *Wright State University*.

medical supplies; and helped mobilize medical staff where needed. The National Cash Register campus was a refuge of hope for people who lost everything and had nowhere else to go.

National Cash Register gave $1 million (the equivalent of $28 million in 2023) to help people recover from the disaster. The company then allocated $600,000 ($17 million in 2023) for a flood control engineering study to prevent such problems in the future.

By 1922, the year of John Patterson's death, the company had sold two million cash registers throughout the world. In 1925, the company decided to go public and issued $55 million in stock ($890 million in 2023), the largest public offering in America's history to that point.

During World War I, the company produced flight instruments for aircraft and a variety of fuses. World War II saw the company manufacturing aircraft engines, bombsights and code-breaking machines, including the American Bombe.

In 1942, German U-boats operating in the Atlantic were attacking and sinking hundreds of American troop and cargo ships bound for Europe.

The problem had to do with the U.S. Navy's inability to break enemy communications regarding their location and targets. To create a decoding machine patterned after a British design, the navy turned to National Cash Register for help. Joseph Raymond Desch (1907–1987), a company electrical engineer, was appointed research director of a special project to develop the navy's version of the Bombe decoding machine capable of breaking the German Enigma machine code. As a result, by 1943, Desch and his extensive staff were laboring in the company's Building 26 and were able to manufacture roughly two Bombe machines per week. The American Bombe was a faster cryptanalytic machine, designed to read communications encrypted by the German Enigma. The company had a staff of highly capable women, known as the Dayton Code Breakers, who did work similar to the British codebreakers at Bletchley Park in England.

Taking full advantage of its former wartime experience with covert communication systems and cryptanalytic equipment, the company was in an excellent position to transition into a major postwar competitor in developing advanced computing and communications technology.

In 1953, company chemists in Dayton invented a pressure-responsive record material to be used as a carbonless copy paper. It was marketed as NCR Paper. Also in 1953, the company acquired the Computer Research Corporation and used it as the foundation to establish a specialized electronics division. In 1956, the production line rolled out its first electronic device, a banking machine based on magnetic stripe technology. In 1957, partnering with the General Electric Company, the company produced its first transistor-based computer, the NCR 304. Five years later, the NCR-315 Electronic Data Processing System was released. It was equipped with the first automated mass storage as an alternative to magnetic tape libraries. In 1968, the company released its first all-integrated circuit computer, the Century 100.

Joseph Desch, seen here about 1960, was placed in charge of the American Bombe development program. The Bombe was to be used by the U.S. Navy as a decoding machine. *National Security Agency.*

National Cash Register officially transitioned to the name NCR Corporation in 1974. Also in 1974, items at a Troy, Ohio grocery store labeled with a Universal Product Code (UPC) were scanned at the checkout with the use of supermarket scanner/computer technology developed by NCR research engineers.

The American Navy Bombe decoding machine, seen here circa 1943, was used against the German navy to crack the codes of their Enigma machine. *Wright State University*.

No longer in the gone-with-the-wind cash register business, NCR is still a major supplier of mainframe computers and approximately one-third of the world's ATM machines. In 2009, NCR announced it was relocating its corporate headquarters from Dayton to Atlanta, Georgia, after Dayton had proudly served as its home for 125 years. But NCR will always be regarded as an Ohio original.

6

BUCKEYE OIL, POWER AND ALLOYS

Materion Brush

Materion Corporation, formerly known as Brush Engineered Materials Inc., is a leading international manufacturer of high-performance engineered materials and the only fully integrated producer of the metal called beryllium. Beryllium is an element that, when combined with other metals like copper, makes an alloy that is lighter than aluminum and with strength closely comparable to that of steel. Most of the applications for beryllium are found in the space aeronautics and defense industries.

Materion's history started with the ingenious Ohio inventor Charles Francis Brush (1849–1929). He was born in 1849 near Cleveland and was the youngest of nine children. Being a youngest child with a passion for science, Brush read a description on how to make arc light. He then focused all of his time toward developing such a device. In high school, Brush succeeded in building an arc light by passing electricity between the gap of two carbon electrodes. While working, Brush spent his off-hours performing experiments to develop an arc light system for commercial use. Such a system would require a powerful and reliable source of electricity. In 1876, with the help of the Telegraph Supply Company, Brush perfected a dynamo to power his arc light and received a patent in 1877. He then engineered an electric arc light with a special regulating system that made such lighting reasonably cost-effective for illuminating streets.

Cleveland, Ohio, took the lead in installing Brush's arc lamps and illuminated its Public Square in 1879. A number of cities followed Cleveland's example in lighting their streets. After the Telegraph Supply Company of Cleveland was restructured in 1880, the name was changed to the Brush Electric Company, which then merged in 1891 with Edison General Electric to later establish General Electric. The creation of General Electric made Charles Brush a wealthy individual and allowed him to pursue other business ventures. He was now in an excellent position to research and develop other inventions that he would later patent, such as the lead storage battery.

Charles Francis Brush, shown here circa 1920, was an inventor who lit the streets of nineteenth-century towns and cities by using an electric arc-lighting system. *Cuyahoga County Public Library.*

In 1926, Charles Francis Brush Jr. (1893–1927), used funding from his father and facilities on the family estate to establish Brush Laboratories with a college friend, Charles Baldwin Sawyer (1894–1964), as his partner. By 1926, Charles Jr. was primarily focused on developing effective industrial uses for beryllium. Unfortunately, in 1927, tragedy overtook the Brush family, when Charles Jr. and his daughter died. They were followed two years later by Brush Sr., who died from pneumonia on June 15, 1929.

Charles Baldwin Sawyer continued to run Brush Laboratories. In 1931, while working with Bengt Kjellgren, he invented a process to extract beryllium from raw ore. The business was then renamed, known as the Brush Beryllium Company. In 1933, an electric beryl furnace was placed into service, and the company slowly began to increase the amount of beryllium metal it could produce, from sixty grams in 1936 to over one thousand grams by 1942.

One of the early major uses for beryllium was in building the first atomic bomb at the U.S. government lab in Oak Ridge, Tennessee. During the postwar years, demand for beryllium continued to grow, especially in the development of nuclear weapons and atomic energy for peacetime commercial uses.

In 1949, Charles Francis Brush III (1923–2006) joined the board of the company. By 1960, sales for Brush had quadrupled due to the high demand for specialty alloys. The company relocated its operations to Elmore, Ohio, where Brush had established its alloy division in 1953.

SCIENTIFIC AMERICAN

[Entered at the Post Office of New York, N. Y., as Second Class Matter. Copyrighted, 1890, by Munn & Co.]

A WEEKLY JOURNAL OF PRACTICAL INFORMATION, ART, SCIENCE, MECHANICS, CHEMISTRY, AND MANUFACTURES.

Vol. LXIII.—No. 25. Established 1845. NEW YORK, DECEMBER 20, 1890. [$3.00 A YEAR. WEEKLY.

1. Windmill in the park. 2. Vertical section of the tower. 3. Dynamo. 4. Storage batteries. 5. Regulating apparatus.

THE WINDMILL DYNAMO AND ELECTRIC LIGHT PLANT OF MR. CHARLES F. BRUSH, CLEVELAND, O.—[See page 389.]

Charles F. Brush constructed a windmill to furnish power for a dynamo that energized an arc-lighting system in Cleveland, as illustrated in *Scientific American* in 1890. *Cuyahoga County Public Library.*

Foundry worker pouring beryllium bronze into ingot molds, circa 1938. *Case Western Reserve University Library.*

The unique properties of beryllium were especially suited in engineering heat shields for missiles, satellites and other aerospace applications. In the 1960s, beryl ore was discovered in Utah's Topaz Mountains, and Brush established a subsidiary, Beryllium Resources, to explore this region and later make a claim on mineral rights. In 1969, the firm decided to become a more independent company by investing in a mining and milling facility in Utah near a mine Brush operated. The mill would process ore into beryllium oxide, which is the raw material for the company's three product lines: beryllium metal, alloys and ceramics.

In August 1971, Brush acquired the S.K. Wellman Division of Abex Corporation. Wellman produced metallic friction material to make sturdy clutches and brakes installed in rugged, heavy-duty, off-road equipment. Brush changed its name to Brush Wellman Inc.

The company focused on diversification and establishing new markets. It completed several acquisitions in the 1970s and 1980s, but the most important emerging new market was Silicon Valley. New companies in Silicon Valley would require heat-resistant beryllium-copper alloy to manufacture microelectronic chips. The higher the heat generated in a microchip's capacity; the more beryllium copper would be required. Brush's beryllia ceramics also came into demand because of the material's superior insulating properties and its ability to dissipate heat.

To participate in the specialty clad metals field and provide materials for power-shift, clutch and brake systems, Brush acquired Technical Materials Inc. in late 1982. In 1984, Brush launched a $57 million capital improvement program, $30 million of which was used for new casting furnaces, rolling equipment and other facilities at the Elmore, Ohio plant, and $15 million of which was invested in constructing the company's first finishing mill in Europe. Also, $10 million was allocated to upgrade the Pennsylvania finishing mill, and $2 million was devoted to warehousing improvements.

New applications for beryllium alloys came in the early 1990s with an increase in electronic technology in the auto industry. The list of new components included automobile relays, connectors, surge protectors, fuse terminals and switches. The company embarked on an expansion program in 1996. The scope of the expansion included new equipment for melting, casting and finishing copper beryllium strips. By the end of the decade, sales grew to a record $456 million (equivalent to $745 million in 2023).

In 2000, Brush Wellman restructured under a new holding company, Brush Engineered Materials Inc. The Wellman operations were liquidated in the 1980s, so this new name reflected the general nature of Brush's products and services. On March 8, 2011, Brush Engineered Materials became Materion Corporation, unifying all of the company's businesses under the Materion name. As communication and transportation industries continue to implement more electronics, there will be a need to rely on beryllium-based components. For this reason, Materion has a bright future.

IDEAL ELECTRIC

In 1903, Stanleigh Glen Vinson established the Ideal Electric & Manufacturing Company in Mansfield, Ohio, out of what had been the Card Electric Company. The original plant, a converted horse carriage shop, occupied 2,500 square feet. Ideal Electric received its first major order from the A. Kieckhefer Elevator Company on May 28, 1903. Ideal developed and produced the first high-torque squirrel-cage motor specifically designed for elevator service. New products also included induction motors and slip-ring motors, introduced as part of the product line in 1905. These motors were engineered for elevators and electroplating. Alternating current motors up to fifty horsepower and direct current motors up to four thousand amperes became part of the product line by 1907. As the company expanded, additional products were manufactured.

During World War I, Ideal Electric supplied many motors for a wide range of applications. Many naval warships were equipped with components fabricated by Ideal Electric. Following the war, Ideal pioneered two major products for the electrical industry. In 1918, it introduced large, slow-speed, synchronous motors and generators using across-the-line starting. These innovations were followed by the polyphase capacitor induction motor.

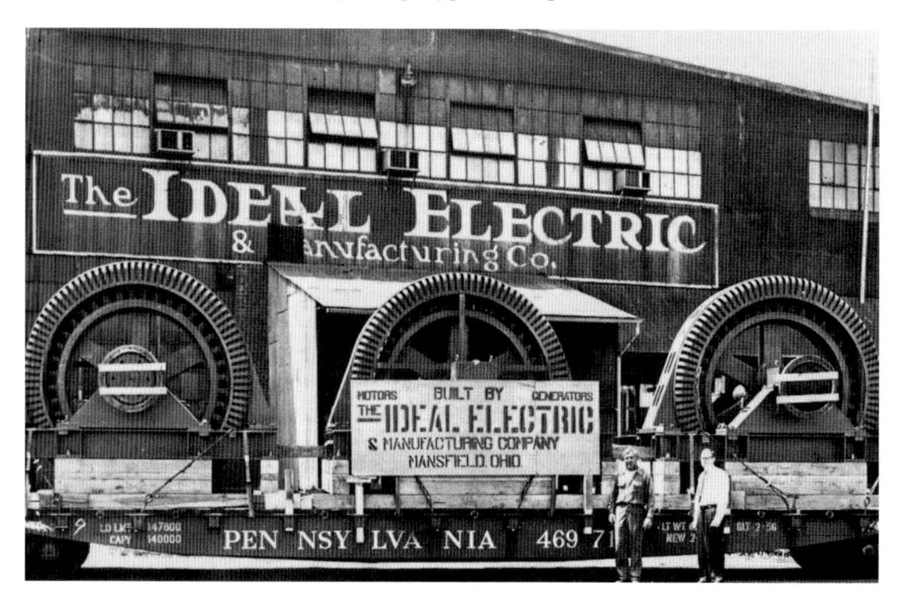

The Ideal Electric & Manufacturing Company of Mansfield, Ohio, prepares to ship a completed order for motors to a customer by rail, circa 1955. *Courtesy Ideal Electric.*

As a result of the postwar increase in business, the company hired highly experienced electrical engineers. Ideal constructed the revolutionary and now historic Mansfield Ideal Ohio Works facility, which sits on twenty-nine acres and has over 280,000 square feet of floor space.

The first official transatlantic telephone call was made on January 7, 1927, when W.S. Gifford, president of the American Telephone & Telegraph Company (AT&T), called Sir Evelyn P. Murray, secretary of the General Post Office of Great Britain. This communication was powered by an Ideal motor-generator set and was the start of the new, intercontinental, across-the-ocean commercial telephone service.

The eddy-current coupling was developed in 1928 to supply variable speed and torque. This coupling is a clutch that acts as a slip device and consists of two rotating elements that are then coupled by a magnetic field. The strength of the magnetic field determines the slip and rotor speed.

During all of World War II (1939–45), Ideal manufactured and delivered generator sets to the U.S. Navy to provide shipboard power for the EC-2 Liberty ships built by industrialist Henry J. Kaiser. These systems also were used to power gun turrets on naval cruisers and supplied power to Allied ground bases throughout the Atlantic and Pacific theaters. Ideal manufactured over four hundred cycle generators for military use during the war.

Ideal engineered a hermetic motor that exceeded the two-thousand-horsepower barrier for use in refrigeration compressors. A hermetic motor used for refrigeration is completely sealed. After a long history of engineering and manufacturing hermetic and semi-hermetic motors for Willis Carrier's inventions, including industrial air conditioners, the ownership of Ideal passed from the Vinson family to the Carrier Corporation in November 1976. Three years later, in June 1979, Ideal Electric was acquired by United Technologies Corporation when it purchased the Carrier Corporation.

On May 28, 1986, Ideal was acquired by a management team under the leadership of Michael M. Vucelic, an engineer and Yugoslavia native who left NASA to go into business for himself. Vucelic had served as the famed NASA Apollo program director credited for the safe return of the *Apollo 13* astronauts. He was also a recipient of the Presidential Medal of Freedom.

In 2001, following the terror attacks in New York City on September 11, Ideal Electric helped the recovery by assisting Verizon in rebuilding major telephone exchanges throughout the city. Two years later, the 2003 Great Northeast Blackout occurred, leaving over forty-five million Americans without electric power. Ideal aided helped supply power to those affected by providing temporary power generation.

Ideal Electric motors equipped with eddy-current couplings were used with heavy-duty pumps to provide drinking water to Columbus, Ohio. Photo circa 1969. *Courtesy Ideal Electric.*

Vucelic's group held ownership until 2007, when Hyundai Heavy Industries of Ulsan, South Korea, purchased Ideal's 280,000-square-foot Mansfield facility. The company was renamed Hyundai Ideal Electric Co. (HIEC). The next year, Hyundai built a 25,000-square-foot assembly floor addition to house machining and welding operations. This modified manufacturing floor also included a new, 14-foot vacuum pressure impregnation tank for motor coils and windings. Ideal continued to lead the industry in 2011 with an expansion that included an additional 30,000 square feet to the assembly and test floor area, which is supported by a new eighty-ton crane with a 35-foot hook height.

Ideal Electric continued to engineer and manufacture quality, custom-built equipment for a variety of applications and customers. Ideal has built small, land-based, high-frequency motor-generator sets for naval aircraft, medium-size generators for naval guided-missile attack vessels, small to large generators for hydroelectric applications, generators for oil platforms, backup power supplies for NASA and small refrigeration motors for buildings like the Houston Astrodome and for the new World Trade Center.

In September 2017, Hyundai sold its entire holdings in Ideal Electric. This returned the company to 100 percent American ownership and set the stage for a resurgence of the factory, led by an entrepreneurial spirit and a passion for electric machinery. Ideal Electric is currently owned and operated by a private American affiliate, making it the only wholly American-owned independent manufacturer of high-power, specialty electric machinery and power systems in the world.

STANDARD OIL COMPANY

The American general public has forgotten that, historically, the nation's oil industry was organized in Cleveland, Ohio, by merchant bookkeeper and Baptist Sunday school teacher John D. Rockefeller (1839–1937). Following Edwin Drake's 1859 success in drilling the first commercial oil well just outside Titusville, Pennsylvania, a young Rockefeller investigated and saw merit in the future of the oil business. Drake's success also initiated the first oil boom in America.

As a careful and studious businessman, Rockefeller observed the gross waste and inefficiencies of small oil companies rushing to take advantage of the new "black gold." His background as a careful bookkeeper therefore gave him an advantage in organizing a company to market oil properly. Rockefeller realized that the real profit was not in drilling for oil but in refining oil into a wide range of marketable products, such as heating oil and kerosene for lighting.

At the height of the Civil War, in 1863, Rockefeller, along with his partners Maurice B. Clark and Samuel Andrews (1836–1904), a chemist, established the Standard Oil Company as an oil-refining business in Cleveland. By 1865, Rockefeller had bought out Clark's full interest. In 1867, he considered asking Henry M. Flagler (1830–1913) to become a partner and to help secure financing for his new expanded venture. Flagler had been working for the Harkness Grain Company, one of several businesses owned by his stepbrother Stephen V. Harkness (1818–1888). Harkness, who originally was a harness maker, established a successful spirits distillery in Monroeville, Ohio, in 1855. In 1864, he was in a partnership to provide crude oil to refineries. Harkness reaped such wealth from the early supply side of the oil business that, in 1866, he liquidated all of his Monroeville holdings and relocated to Cleveland's Millionaires' Row on Euclid Avenue. In 1867, using his personal wealth for Rockefeller's venture, Harkness invested $100,000 (equivalent to $2.1 million in 2023). Harkness stipulated that Flagler be installed as a full partner in control of Harkness's full interest, because Harkness wanted to remain a silent partner.

John D. Rockefeller was a reserved man and one of the original founding partners of the Standard Oil Company of Cleveland, Ohio. He is regarded as America's first billionaire. Photograph circa 1880. *Case Western Reserve University Library.*

In 1870, John D. Rockefeller, Henry M. Flagler, Samuel Andrews, Stephen V. Harkness and William Rockefeller signed the Articles of Incorporation that became the Standard Oil Company, Incorporated. Standard Oil was now operating the largest petroleum refineries in Cleveland. Over the next decade, John D. Rockefeller focused on eliminating inefficient competitors by acquiring them and then shutting them down. He also formed favorable mergers and negotiated to obtain railroad freight transport rebates. For example, in 1868, the Lake Shore Railroad, part of the New York Central, gave Rockefeller's company a 71 percent discount rate provided that at least 60 carloads of oil barrels were received each day and that the Standard Oil staff would load and unload it. This deal was based on volume, which smaller refineries could not produce to qualify for such discounts. Between 1865 and 1870, Standard Oil's clandestine transportation arrangements reduced the price of kerosene from fifty-eight cents to twenty-six cents per gallon. By 1880, Standard Oil of Cleveland, Ohio, controlled 95 percent of petroleum refining in America.

In 1882, Rockefeller and nine additional trustees signed the Standard Oil Trust Agreement. This document stipulated that all affiliated companies specified as part of the agreement could be purchased, created, dissolved or divided. The agreement was cleverly crafted by Standard Oil attorneys with the intent of discouraging any public investigation into the actions and proceedings of the trust. For years, the heavy layers of legal paperwork and loopholes protected the trust. Eventually, the trustees governed nearly forty corporations, including fourteen wholly owned operations. In 1892, the Ohio Supreme Court attempted, to no avail, to dissolve the trust by a straightforward court order. But in 1885, the Standard Oil Company relocated its headquarters to New York City, where it effectively continued to manage the trust as usual.

From 1882 to 1906, Standard Oil paid stockholder dividends amounting to $548,436,000, a 65.4 percent payout ratio, given that the total net earnings for that same duration amounted to $838,783,800 ($27.2 billion in 2023).

Using the Sherman Antitrust Act of 1890, the U.S. federal government managed, following a two-year trial, to prove that the Standard Oil Company was a monopoly. But it required testimony from nearly four hundred witnesses and the presentation of over 1,300 exhibits and a 1,500-page report.

As chairman of Standard Oil, Rockefeller governed the company until he retired in 1897, although he continued to be the company's majority shareholder. At the conclusion of the antitrust hearing in 1911, it was ordered that the Standard Oil Trust be dissolved into thirty-four smaller companies. Rockefeller continued to hold stock in all of the smaller companies, which only increased in value. At the time of his death, his net worth was $1.4 billion (equivalent to $28 billion in 2023), making him the wealthiest person of the twentieth century.

Despite the breakup of the original Standard Oil Trust into smaller companies, the largest of these companies remain the foundation of the American oil industry in the twenty-first century. Standard Oil of New Jersey became Exxon Oil, and Standard Oil of New York became Mobil Oil. Both were later combined to form ExxonMobil. Standard Oil of Kentucky, Texaco and Unocal are today known as Chevron. The Ohio Oil Company became known as Marathon Oil and later gave birth to a separate company, Marathon Petroleum, which is headquartered in Findley, Ohio. Standard Oil of Indiana became AMOCO. Some people may still recall Standard Oil of Ohio, which was better known as SOHIO. Both AMOCO and SOHIO were acquired by British Petroleum (BP).

One last note about Henry M. Flagler: in the state of Florida, he is best remembered as the pioneer founder of the city of Miami and Palm Beach.

7

RUBBER AND THE EVOLUTION OF THE SMOOTH RIDE

Timken

Canton, Ohio, is home to the Timken Company, the world's largest manufacturer of tapered roller bearings. This mechanical component is composed of a "tapered cup" and a "tapered cone" and is one of the basic designs for the bearing industry. Tapered roller bearings are more common and versatile than any other bearings and are standard components in everything from trucks, automobiles, agricultural harvesting equipment and railroad equipment to material conveying lines.

The design of these bearings is based on an improved tapered roller bearing patented in 1898 by Henry Timken (1831–1909). Timken began his industrial manufacturing career as a carriage maker in St. Louis in 1855. Years later, he was familiar with the difficulty that heavy wagons had in executing sharp turns. In order to solve the problem, he conceived a tapered roller bearing that was capable of handling both radial and thrust loads induced when cornering. The positive results provided a number of benefits. First, a fully loaded wagon operated more smoothly. This reduced wear on components and lowered replacement costs. The bearings improved the general performance of the wagon, and in some cases fewer draft animals were needed to pull a loaded wagon. In addition, smoother cornering reduced the chance of losing a load of trade goods. In 1877, Timken received a patent for the Timken Buggy Spring. This was the first of his thirteen patents. The spring design was used widely throughout the country and was licensed to manufacture by several other companies. With a growing carriage business and the success of the buggy spring, the Timken name became well known across the country.

Henry Timken's technological advances would later extend to the automobile industry and reduce needless wear and tear on practically all motor vehicles. Ball bearings were already in use when Timken's two sons, Henry H. and William Timken, worked alongside their father in his carriage-making shop.

In 1895, Henry Timken challenged his sons to come up with an improvement that would reduce friction and improve the reliability and longevity of the axles on their carriages. After three years of experimenting, Henry H. and William Timken and their father patented their tapered roller bearing in 1898. The elder Henry Timken's foresight paved the way for the Timken Company to make the transition into serving the automotive and trucking industries. The next year, they established the Timken Roller Bearing Axle Company.

Henry H. Timken (1868–1940) deliberately began to call on car pioneers in their fabrication shops in order to learn about the needs and requirements of their vehicles. He used this knowledge to design and manufacture friction-reducing bearings that satisfied the performance requirements of a wide range of vehicles. The strong desire of Henry H. Timken to create high-quality products is what ultimately made the company the largest manufacturer of bearings in the world. In 1901, the Timken's decided to relocate the Timken Roller Bearing Axle Company to Canton, Ohio, because of its proximity to the auto industry in Detroit, Cleveland and Dayton as well as the steel mills in Pittsburgh and Cleveland.

The demand for Timken axles and bearings grew significantly after Henry Ford implemented assembly-line methods in the production of the Model T in 1908. In 1909, the Timken brothers separated the axle division and relocated it to Detroit. This move launched the new Timken-Detroit Axle Company with William Timken (1866–1949) as president. By 1909, the same year Henry Timken died, the company output was well over 855,000 bearings annually, and it had a payroll of approximately 1,200 employees.

In 1915, Timken started producing its own steel to ensure an adequate supply on hand for its manufacturing operations, despite World War I material shortages. That year, the company began constructing a steel tube mill and smelt foundry at its Canton location. With the ability to produce its own steel, Timken became the first bearing manufacturer to be its own steel

An extremely large industrial bearing assembly, circa 1918. The tapered roller bearing was first patented in 1898 by Henry Timken. Stark Library.

supplier. In 1919, a separate Industrial Division was organized to replace the company's Farm Implement and Tractor Division. The Industrial Division's goal was to develop bearings for a wide variety of industrial uses, including electric motors, concrete mixers, printing presses and elevators.

The market for Timken bearings and steel expanded rapidly throughout the 1920s. In 1920, the company opened a bearing plant on Fifth Avenue in Columbus. This was the first facility outside of Canton. Being conscious of the impact its operation had on the environment, that same year, the company built a wastewater treatment plant to service the Canton facility. In 1922, Timken stock was offered to the public, and the company opened an assembly plant in Canada that year.

The Timken steel foundry floor in Canton, Ohio, circa 1910. *Stark Library*.

Timken bearings were introduced to the railroad industry in 1923, when bearings specifically designed by Timken were first tested on an intercity streetcar run from Canton and Cleveland. Later that year, the bearings were installed and tested on a railroad boxcar. By 1926, other railroad companies had gained the understanding that the tapered bearings allowed a train to travel safely at faster speeds. A large order was placed by the Chicago, St. Paul & Pacific Railroad for use on high-speed passenger trains.

One Timken subsidiary is the MPB Corporation, which produces super-precision and miniature bearings used in applications ranging from missile-guidance systems to computer disk drives. While most Timken facilities are located in the United States, it is a multinational corporation with manufacturing operations in Taiwan, India, Sweden, Romania and other countries.

It should be noted that on September 19, 1998, in Akron, Ohio, the elder Henry Timken was posthumously honored by being inducted into the National Inventors Hall of Fame. Throughout the twentieth century and into the next millennium, the Timken Company continues to be a world leader in the manufacture of bearings and steel. It is committed to providing a full range of industrial motion needs with a focus on growing as a power transmission leader.

B.F. GOODRICH

Early American industry was grounded in Akron, Ohio, better known as the "Rubber Capital of the World" because it was the main headquarters of four major tire companies established there. The local industry began with B.F. Goodrich Corporation in 1869. Nearly three decades later, the Goodyear Tire and Rubber Company began operations. This was followed by Firestone Tire and Rubber Company two years later in 1900 and General Tire in 1915. The numerous jobs the rubber factories provided for deaf people led to Akron being nicknamed the "Crossroads of the Deaf."

Benjamin Franklin Goodrich (1841–1888) was one of the first industrialists to plant the rubber industry in Akron. His career began as a physician. He served during the Civil War as a rank captain and battlefield surgeon in the Union army. After the war, Goodrich retired from medicine and went to work in the booming Pennsylvania oil fields and on occasion speculated in land development.

In 1869, Goodrich entered into a licensing agreement with Charles Goodyear and acquired the Hudson River Rubber Company with his partner, J.P. Morris. The company was originally located in Melrose, New York, but it quickly folded. The following year, Goodrich received and accepted an offer from the community of Akron to relocate his business there in exchange for $13,600 in startup capital (equal to $302,000 in 2023). With the funding from the local citizens, he and his brother-in-law Harvey W. Tew established Goodrich, Tew & Company. But in a short time, Goodrich bought out Tew's entire holdings.

Benjamin Franklin Goodrich, seen here about 1882, was a physician before getting into the rubber business. *Akron-Summit County Library.*

At the beginning of the 1870s, there were no rubber manufacturers in America west of the Allegheny and Cumberland Mountain ranges. Goodrich's expectation was to dominate the rubber industry in the Midwest and beyond the Mississippi River. He established his Akron factory, known as the Akron Rubber Works, in March 1871. With an initial staff of twenty employees, Goodrich produced a variety of rubber items but concentrated on manufacturing firefighting hoses that would not fail under pressure, unlike regular leather hoses, which cracked at low temperatures. Goodrich solved this problem by encasing the rubber hose in a heavy cotton fabric that would reinforce the rubber under pressure and resist freezing. In time, Goodrich helped alleviate the use of a bucket to water home gardens by marketing garden hoses.

Given the late nineteenth-century demand for bikes, the company also manufactured rubber bicycle tires. In 1880, the company was formally renamed the B.F. Goodrich Company. Growth was minimal during the first decade of operation. On at least two occasions, the firm came close to declaring bankruptcy. The company's future was still uncertain at the time of Goodrich's death at the age of forty-six in 1888. Also in 1888, John Boyd Dunlop (1840–1921), a Scottish-born veterinarian and inventor living in Ireland, invented the first practical pneumatic tire for use in bicycle racing. Pneumatic tires are made of rubber and inflated with air. He sold his rights to the pneumatic tire and, in time, a company that bore his name was incorporated: the Dunlop Tire Company. It was headquartered in Akron. Because it made for a smoother, more comfortable ride, Dunlop's pneumatic tire was extremely popular with bicyclists.

The demand for tires grew exponentially following the development of the automobile. The first automobile tires were solid rubber, similar to those used on carriages. The B.F. Goodrich Company went to work engineering a pneumatic tire that was practical for automobiles and able to endure the speeds and loads of various vehicle applications. By 1892, just four years after the death of B.F. Goodrich, the company was employing a staff of over four hundred workers and was generating sales exceeding $1.4 million ($44 million in 2023). In 1903, the first automobile to cross the continental United States was fitted with Goodrich tires. By 1911, the company's gross sales exceeded $27 million ($812 million in 2023).

The innovative chemists and engineers of B.F. Goodrich Company developed plasticized polyvinyl chloride (PVC) in 1926 and synthesized rubber in 1937. In 1927, pioneer aviator Charles Lindbergh made his nonstop solo transatlantic flight with his *Spirit of St. Louis* aircraft, equipped with Goodrich tires. The company also created the first tubeless tire in 1946. In 1961, the National Aeronautics and Space Agency (NASA) asked

Goodrich to develop the first spacesuits for its astronauts. The company's tire brand name was acquired by Michelin Tire in 1988, and Goodrich completely bowed out of the tire business and turned its attention to the aerospace industry. In 2001, the B.F. Goodrich Company changed its name to the Goodrich Corporation. It is regarded as an international supplier of systems and services to the aerospace and defense markets. The company eventually relocated its headquarters from Akron to Charlotte, North Carolina. In 2012, the company was acquired by United Technologies.

SYNTHETIC RUBBER

In 1937, B.F. Goodrich chemists created a cost-effective synthetic rubber. This proved to be timely, because during World War II, the nation experienced a rubber shortage. Asia was the source of most rubber imported to the United States, and Japan controlled much of Asia and stopped all rubber exports to the United States. The U.S. military needed a lot of rubber for the war effort,

Synthetic rubber coming off the rolling mill at B.F. Goodrich Co. and ready for drying, circa 1942. *Library of Congress.*

especially for jeeps, trucks and aircraft tires. The company also created the first tubeless tire in 1946.

By the start of World War II, the cost to manufacture synthetic rubber had dramatically declined. In Akron, Goodrich and other tire companies, including Goodyear Tire and Rubber Company and Firestone Tire and Rubber Company, took the lead in manufacturing a wide range of synthetic rubber products, ending the country's dependence on foreign rubber. Following the end of the war, these Akron rubber companies began turning out synthetic rubber products for peacetime use.

GOODYEAR TIRE AND RUBBER

Another Akron original is the Goodyear Tire and Rubber Company. The company was started by Frank Seiberling (1859–1955) in 1898 with $3,500 (equivalent to $120,000 in 2023). In 1906, Seiberling was installed as the company's fourth president, and the company had over 350 employees. That year saw the company start manufacturing the world's first quick, detachable, straight-side tire. Additionally, the Seiberling-Stevens patent was issued for a tire-building machine.

Because of the demand for vulcanized rubber products, the company soon emerged as a leader in the manufacture of pneumatic bicycle and automobile tires and a wide range of rubber hoses. Goodyear played a key role in the innovative development of new rubber products that expanded the industry. In 1903, the company's research efforts led to Paul Litchfield, who would later become Goodyear's first CEO, being granted a patent for the first tubeless automobile tire.

The year 1907 was a pivotal one for the company, starting with Henry Ford placing an order for 1,200 sets of tires for his Model T cars. As a result of this, a night shift was implemented. Also in 1907, Goodyear started marketing detachable rims directly to auto manufacturers and issued competitors a license to manufacture both detachable rims and tires. By 1909, the company developed the first airplane tires for use in the aviation industry. Before this, pilots primarily used bicycle tires on their planes. Goodyear manufactured its first blimp in 1912 and developed a bulletproof fuel tank for airplanes in 1919. The company later established a division to build helium-filled airships. It still maintains a massive hangar facility in the Akron area.

This is a circa 1910 interior look at the Goodyear Tire and Rubber Company plant, where the tire casings and tread belts were assembled prior to vulcanization. *Akron-Summit County Library*.

Goodyear manufactured its first blimp in 1912 and later established a division to build helium-filled airships. Photograph circa 1928. *Akron-Summit County Library*.

By 1926, Akron, Ohio's Goodyear Tire and Rubber Company was ranked the world's largest rubber manufacturer, surpassing B.F. Goodrich. The company remains an industry leader in creating innovative products for the twenty-first century. In 2021, the Goodyear Tire and Rubber Company sold nearly $17.5 billion worth of products. The firm employed more than seventy-two thousand workers and is still ranked internationally as the largest tire manufacturer.

Seiberling remained the company's president until the economic recession of 1921, prompting he and his brother Charles Seiberling (1861–1946) to resign. Soon thereafter, the Seiberling brothers established the Seiberling Rubber Company in Barberton, Ohio.

FIRESTONE TIRE AND RUBBER

Firestone Tire was founded by Columbiana, Ohio native Harvey Samuel Firestone (1868–1938). He attended the Spencerian Business College in Cleveland prior to being employed as a bookkeeper for a coal company in Columbus. While in Columbus, Firestone went to work as a salesman with the Columbus Buggy Company but was later laid off following a business downturn after the panic of 1893. Finding himself unemployed, Firestone decided to put his previous experience to work with the innovative idea of retrofitting steel-rim wheels on buggies with solid rubber tires. He knew that a rubber tire would make for a more comfortable ride. Firestone was clearly focused on manufacturing and selling the rubber tires.

In 1895, his next move was to Chicago, Illinois, where he purchased a factory. With only one employee, Firestone started the Firestone-Victor Rubber Company. Within a few months he had changed the factory's name to the Firestone Rubber Tire Company. As a result of Firestone's sales experience, his rubber tire business quickly realized a profit. After being in business for just four years, Firestone sold the company for a $45,000 profit ($1.5 million in 2023).

In 1900, Firestone moved to Akron, where he established the Firestone Tire and Rubber Company. Initially, he relied on outside suppliers to fill his need for rubber. His firm simply fastened the rubber to steel carriage wheels. In its first year of operation, the Firestone Tire and Rubber Company grossed more than $100,000 ($3.4 million in 2023). Within three years, the company had begun to manufacture rubber from unprocessed raw rubber.

Henry Ford (*left*) and Harvey Firestone had a special working relationship that complemented both men as leaders in the transportation industry. They are seen here circa 1919. *Wayne State University Library*.

By 1904, it started developing and subsequently manufacturing pneumatic tires for the new automobile industry.

Henry Ford was interested in using pneumatic tires on his mass-produced automobiles. He met with Firestone to discuss the matter, and Firestone assured him of the benefits of air-filled tires. After this conversation in 1905, Henry Ford decided to place his first order for tires from Firestone. Harvey Firestone immediately increased his staff, from 12 to 130 full-time employees. In 1906, the Firestone Tire and Rubber Company manufactured and sold over twenty-eight thousand tires, with total sales exceeding $1 million ($34 million in 2023). By 1910, the company was turning out over one million tires annually. Firestone's engineered innovations in tire design allowed vehicles to travel more safely at greater speeds and longer distances.

Firestone took an interest in providing tires engineered especially for automobile racing. The 1911 inaugural race of the Indianapolis 500 was won by a driver equipped with Firestone tires. Between the years 1920 and 1966, every car that won the Indy 500 was equipped with Firestone tires.

Harvey Firestone remained as president of the Firestone Tire and Rubber Company until 1932, when he retired from the firm's active management.

He remained chairman of the board of directors. Firestone passed away in 1938, leaving a legacy of helping make Akron, Ohio, the rubber capital of the world. In 1988, a Japanese company, the Bridgestone Corporation, acquired Firestone.

While the Firestone Tire and Rubber Company prospered, its workers sometimes suffered. In the late nineteenth and early twentieth centuries, factory employees faced poor working conditions, low wages and almost no benefits. This was true for those employed by rubber manufacturers in Akron, including Goodyear Tire and Rubber, B.F. Goodrich and Firestone. In an attempt to alleviate their conditions, workers established a union, the United Rubber Workers, in 1935. The following year, this union organized its first major strike in Akron's rubber industry.

The strike began as a protest against a plan created by Goodyear to reduce wages and increase the pace of production. The workers utilized the concept of the "sit-down" strike. In the past, when workers went on strike, they left the factory to join picket lines. Company owners often hired "scab" laborers to cross the picket lines and continue production. This practice made it difficult for striking workers to obtain their demands. In contrast, in a sit-

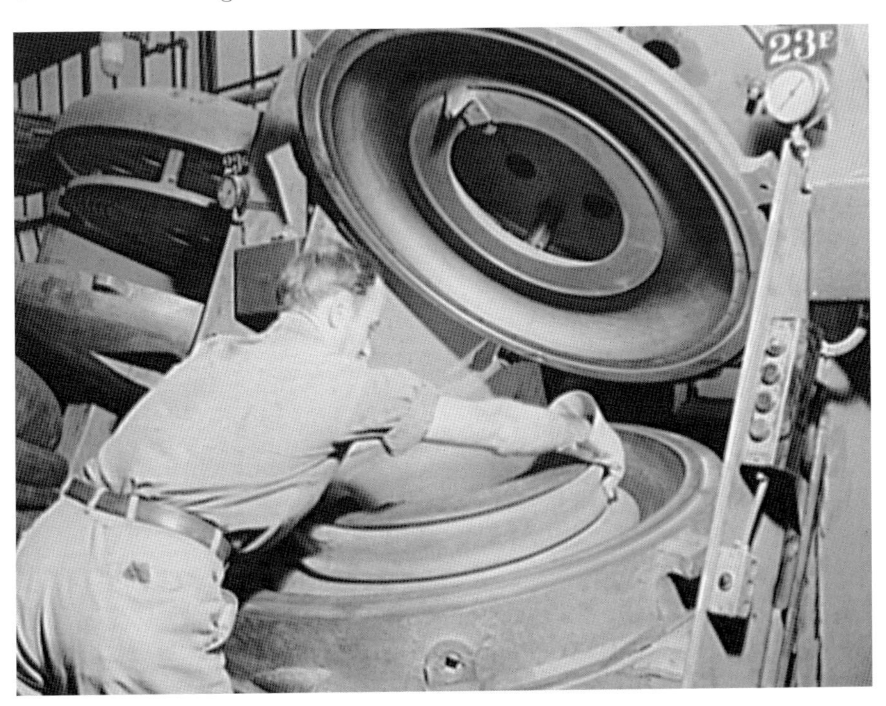

This skilled worker is building a tire prior to its being vulcanized, circa 1932. *Library of Congress.*

These women are working in the Firestone division making spark plugs, circa 1938. *Library of Congress.*

down strike, employees stopped working but still occupied their places in the factory. This meant that factory owners could not send in additional workers to continue the job. In the long term, the rubber companies were forced to recognize the United Rubber Workers and negotiate better contracts. One immediate success was a six-hour workday.

Several reasons existed for the workers' success in this strike. First, sit-down strikes made it much more difficult for employers to replace striking workers. Equally as important in this strike was the federal government's recent passage of the Wagner Act. This legislation made unions legal for the first time in U.S. history. Finally, the United Rubber Workers belonged to a larger organization, the Congress of Industrial Organizations (CIO). The CIO consisted of an umbrella organization for multiple unions. These unions worked together, providing both moral and material support to CIO-member unions, especially when these unions went on strike.

In 1988, the Bridgestone Corporation purchased Firestone. Bridgestone is involved in tire manufacturing on an international scale, making it one of the largest of its kind in the world.

8

MILLING MACHINES, TRAIN TRAVEL AND FASTER FLIGHT

Cincinnati Milling Machine

In the twentieth century, the Cincinnati Milling Machine Company was one of America's best manufacturers of high-quality machine tools. The company's story began in 1874, when George Mueller formed a partnership with Fred Holz after Mueller inherited his father's machine shop in Cincinnati. They continued manufacturing parts for sewing machines and repairing small machines like Mueller's father had done, but they also began building specialized machinery, including a device that produced screws and the taps and dies that could cut threads. By 1876, the company focused mostly on manufacturing the machines that made screws. Mueller and Holz renamed the firm the Cincinnati Screw and Tap Company.

Around 1878, the company couldn't afford a new milling machine for the manufacture of taps, so Holz used his machinist skills to fabricate an improved basic milling machine. Other shops also needed such a machine, so, by 1884, the company had incorporated to raise the capital needed to expand its operation and relocate near the Ohio River.

In 1887, while doing business with the Cincinnati Screw and Tap Company, twenty-one-year-old Frederick Augustus Geier (1866–1934) purchased a stake in the company. While Holz focused on the engineering side of the business, Geier put his time into sales and organization. He favored concentrating on manufacturing milling machines instead of the company's screw and tapping machines. In early 1889, the screw and tap business was sold to a group of employees. Mueller, Holz and Geier remained with the milling machine business and renamed it the Cincinnati Milling Machine Company, or the Mill, as many referred to it. Two years later, cofounder Mueller sold his interest in the company.

Although the milling machine industry had long been controlled by companies on the East Coast, the Mill found its niche by emphasizing quality. In 1889, Holz engineered a cutter grinder that would save money by sharpening cutting tools, enabling them to last longer. This was a new product to sell with its mills and helped to build the company's industrial machine tooling reputation.

Holz retired and sold his entire holdings in the company to Geier. Geier purchased land in 1907 and started construction on an expanded foundry and power plant. This new complex was placed in service in 1911 with over six acres of floor space. It was the largest milling machine factory in the world. Geier foresaw the popularity of automobiles and the birth of a metal-driven industry that would require dependable and accurate milling machines.

After two years of World War I in Europe, the Allies looked to America for machine tools to gear up for war production. By 1916, the federal government had initiated a war-readiness program, which only increased the demand for machine tool makers. Sales for the Mill were $1.4 million in 1914 and grew to $7 million in 1917 ($155 million in 2023). The labor force increased from 310 to 1,270 workers.

With the end of World War I and government contract cancellations, the Mill found that it suddenly had an excess inventory of milling machines. Salaries and general wages were reduced by 15 percent, and staffing decreased from approximately 1,004 in 1920 to 250 by 1921.

Geier's son Frederick V. Geier (1894–1981) suggested that the company add a new line of center-type grinding machines. This was a device that used wheels to grind metal into round or cylindrical precision parts, such as bearings, pistons and valve stems. This move would more fully utilize the plant's fabrication capability. In September 1921, the company acquired controlling interest in the Cincinnati Grinder Company. By 1926, the success of this expansion made Cincinnati Milling the nation's largest machine tool company.

During the Great Depression, the Mill's exports grew to represent one-third of sales in 1932 and helped sustain the company during this difficult time. Prior to the Depression, the company used sales representatives and distributors to sell to customers. But in 1931, the company worked with the

Nos. 2 and 3 Plain High-Power Cincinnati Millers

This is a World War I–era Cincinnati horizontal milling machine, driven by belts from an overhead line shaft, circa 1917. *Cincinnati & Hamilton County Library.*

University of Cincinnati to develop a co-op program that provided sales training for students. After students' graduation, the Mill was able to place its own sales staff in the field and started selling directly to customers.

In the 1930s, the Mill introduced hydraulic-powered machines, broaching machines that made finer and more precise cuts and the dial-type milling machine, with a power-speed and feed-change feature that became an industry standard.

In 1934, the elder Geier passed away and his son Frederick was elected president of the company. Frederick's first project was to establish a subsidiary in Great Britain to manufacture machines for the international market. During World War II, in 1940, Britain purchased milling machines from the Mill amounting to $16.5 million ($331 million in 2023). The Mill had foreseen this need, expanding the company's capacity in 1938 and initiating an in-depth training program for its staff, which was increased to 8,600 employees during wartime.

In 1943, Geier and his executive staff focused on strategic planning for postwar growth and identified four main goals. These were as follows: to develop new technologies, broaden machinery product lines, extend overseas manufacturing and enter the industrial consumables market. With these goals in mind, the Mill went public in February 1946 to fund new undertakings.

To broaden the line of machinery products, the company developed new technologies for making lathes, grinding wheels and cutting fluids. In 1955, the Mill acquired a chemical company that made additives for plastic and, in 1957, started working with glass-reinforced plastic components. As the 1950s came to a close, the Mill began using a new electronic technology called numerical control, which used punch cards with numerically coded instructions. This technology enabled a computer to control the movements of machine tools and was the start of industrial automation.

By the end of the 1960s, the Cincinnati Milling Company had evolved into a more diverse business than its name designated. In May 1970, the stockholders approved the name Cincinnati Milacron, because Cincinnati is a city associated with the machine tool industry and the new term *Milacron*'s root meaning is "highest precision." Company staff continued to use the name the Mill.

The 1970s was a time of streamlining and transition and saw both Frederick V. Geier and Philip O. Geier Jr. retire from their executive positions, leaving James A.D. Geier in charge as chairperson. Under James's leadership, the company focused on computer-control technology for its machine tool equipment and established a research department to develop non–machine tool products. As a result of this research, the Mill moved into the plastics machinery business and manufactured its first plastics injection molding machine in 1968.

The late 1970s saw the Mill become one of the world's first manufacturers of computer-controlled industrial robots. These robots, which performed such tasks as spot-welding, became integral to assembly lines at companies like General Motors, Ford, Volvo and General Electric.

In 1980, the Mill focused on its three main divisions: machine tools, industrial specialty products and plastics machinery. During the 1980s, the

A motor-driven horizontal milling machine, a vital piece of equipment in any tool and die shop. This machine is shown circa 1942. *Library of Congress.*

company felt the severe effects of Japanese competition and began cost-cutting measures, including plant consolidations and early retirement plans. The company's worldwide employment staff was reduced by 35 percent. As a result, in 1983, the company reported its first annual loss and continued to report losses in five of the next ten years. Increasingly tough competition and misdirected business turns took their toll on the Mill.

A big transition took place in 1998, when the company divested itself of more than a century in the machine tool industry and modified its name to Milacron Inc. Today, Milacron is a company that manufactures and distributes plastic-processing equipment for fields such as injection molding and extrusion molding and is now a subsidiary of Hillenbrand Inc., which acquired Milacron in November 2019.

BARNEY & SMITH RAILCAR BUILDERS

In 1849, Eliam Eliakim Barney (1807–1880) and Ebenezer T. Thresher (1799–1886) formed a partnership in Dayton, Ohio, and established the Dayton Car Works to manufacture streetcars for horse-drawn railways.

Thresher and Barney collaborated with Elijah Packard, an experienced railcar builder, and bought a tract of land and erected a railcar fabrication building. The new venture was called the Thresher, Packard & Company. The firm began at a time when Dayton had no railroad tracks, so shipping railcars could be done only by using freight canal boats.

Within a few years, Elijah Packard, who was the only experienced railcar builder, had passed away. But he had assembled a skilled staff capable of continuing the work in a competent manner. The business was renamed E. Thresher & Company. As Ebenezer Thresher's health started to fail, Barney became senior manager of the company. In 1854, he took on a new partner, Caleb Parker, and the firm was reorganized as the Barney, Parker & Company. In 1858, Thresher sold his holdings to Barney and Parker. Parker had been a partner in Dean, Packard & Mills Car Manufacturers of Springfield, Massachusetts. This was the same railcar business Elijah Packard had been associated with. Parker shut down all operations of that business and relocated its full inventory of equipment and machinery to Dayton.

In 1855, the company turned out its first sleeping cars. They left the factory ready for service and outfitted with blankets, quilts and pillowcases. Barney had a reputation for quality and craftsmanship, but the railcar output was another thing entirely. To make full use of the factory's capability, that same year, Barney acted as a subcontractor to Cyrus H. McCormick to manufacture four hundred reapers. During the panic of 1857, the company went bankrupt and had to close its doors for nearly a year. Fortunately, by 1859, the market was rebounding. Barney, Parker & Company now employed 1,250 workers and was manufacturing passenger railcars and sleeping cars for private individuals who operated them on railroads in southern states.

The Barney & Smith Manufacturing Company in Dayton, Ohio, seen here about 1889, made quality railroad coaches and sleeping cars. *Dayton and Montgomery County Library.*

The Civil War brought a recession for many, but railcars for the war effort were in demand in the North. By the middle of the war, the Dayton-based railcar manufacturer was working at full capacity and producing almost exclusively for the military. Following the war, the company was a firmly established rail equipment manufacturer. Without any outside help, the original $10,000 capitalization had been increased to $500,000 ($9.7 million in 2023). The factory was expanded to turn out twenty freight cars weekly and two passenger cars each month.

Caleb Parker decided to retire in 1864 and sold his share of the company to Preserved Smith (1820–1887). Smith, who was a Presbyterian, had served as a railroad-line financial manager prior to purchasing Parker's interest. In 1867, the company was restructured as a joint stock firm and was now known as the Barney & Smith Manufacturing Company. During the panic of 1873, the company fared well financially and garnered additional business by manufacturing horse-drawn railcars for local street transportation. In 1877, the company started manufacturing some of the first circus cars used to haul wagons, equipment and performers. Sleeping car production was expanded, and the company entered the new narrow-gauge track market. By 1878, Barney & Smith was one of the few railcar manufacturers in the country able to accommodate standard-gauge, broad-gauge and narrow-gauge railroads.

Eliam Barney was active in the company until his death in 1880. He served as the lead salesman, supervised material and lumberyard inventory and gave attention to all kinds of new developments in the railcar manufacturing industry. Barney & Smith had been the nation's largest railcar manufacturer until George Pullman opened his factory town near Chicago in 1881. Barney & Smith employed over 1,500 workers at this time, and in 1882, capital stock in the firm increased to $1 million ($30 million in 2023). These funds would be for capital improvements and additional growth.

By 1890, Barney & Smith employed over two thousand workers and entered the field of electric traction railcars. Its railcars were in service in Ohio, Indiana, Illinois and Pennsylvania. The company grew to be a major manufacturer of both passenger and freight railcars in the latter half of the nineteenth century. Unfortunately, Barney & Smith suffered from a lack of forward thinking and began to stagnate. As a manufacturing business, it held no patents, unlike Pullman, which was leading the industry in patents for railway passenger cars.

In 1892, the Barney family wisely sold its entire interest in the company to a group of Cincinnati investors for $4.5 million ($144 million in 2023). But the new owners of Barney & Smith had their expectations for sizable gains

A special-order circus advance railcar for the John Robinson Circus, circa 1895. *Dayton and Montgomery County Library.*

quickly vanish due to the panic of 1893. By the end of 1897, freight car manufacturing was increasing in orders, and there was a particular demand for the new steel car. Barney & Smith decided to take action in 1905, but by then, 75 percent of the steel car market windfall was gone.

By 1912, the markets for interurban railcars and narrow-gauge trains were gone, and Barney & Smith management lacked a sense of direction. Then, on Easter weekend, the Great Flood of 1913 devastated most of the Midwest, including downtown Dayton. The factory complex was so severely damaged by the flood that the company went into receivership and struggled to survive. By the time it came out of receivership in 1915, there were barely enough orders to sustain business. The federal government seized the railroads in 1917 for the war effort. During World War I, the only business was for low bidders, and Barney & Smith was unable to comfortably compete due to their overhead and cost. The postwar recession years reduced orders further. The business that at its peak had over 3,500 workers in a manufacturing complex covering fifty-nine acres closed its doors in February 1921. Barney & Smith Manufacturing Company of Dayton went out of business after seventy-three years as a quality passenger, parlor, sleeping, dining, freight, coal and refrigeration railcar manufacturer.

LIMA ARMY TANK PLANT

The Lima Army Tank Plant is a military tank manufacturing facility located in Lima, Ohio. The plant is owned by the federal government but is currently operated by General Dynamics Land Systems, a private contractor.

With war raging in Europe the U.S. Army began constructing the Lima Army Tank Plant, equipped with a steel foundry, in May 1941 to fabricate centrifugally cast gun tubes. But this method of production was soon rendered obsolete. Next, the army converted the plant to a heavy vehicle logistics facility to process combat tanks for overseas and domestic shipping. In November 1942, the General Motors subsidiary United Motors Services was issued a government contract and assumed full operation of the plant to manufacture tanks for the American and Allied war effort. The plant made many vehicles, including the M5 Stuart tank (1941–44) and the M26 Pershing tank (1943–45) for the European theater.

Following World War II, the plant was designated the Lima Ordnance Depot and was used as a receiving and long-term warehousing facility for

The M26 Pershing tank was made in Lima, Ohio, from 1943 to 1945 during World War II and saw action in both theaters of the war. Photograph taken circa 1942. *Library of Congress.*

combat vehicles no longer in service. With the rise of the Korean conflict, the plant updated and made technology modifications on tanks prior to being prepared for shipment. Other tanks produced at the Lima facility include the M1 Abrams (1980–86).

Immediately following the Vietnam War, the army decided in favor of a Chrysler-engineered combat vehicle. This would become the Abrams tank, named in honor of General Creighton Abrams (retired). At the time, Chrysler was the Lima plant's operations contractor and in charge of initially manufacturing the M1 tank. The first round of the tank was completed in 1980. This was a time when Chrysler as an automobile manufacturer was in financial turmoil. In order to protect the tank contract from possible bankruptcy, a separate subsidiary, Chrysler Defense, was put in place as the plant operations contractor.

Chrysler Defense was turning out thirty Abrams per month, but in 1982, the decision was made by Chrysler to sell the defense subsidiary to General Dynamics. In August 1985, the plant began manufacturing the new M1A1 model tank and eventually was turning out 120 of these tanks per month. On closing its Detroit Arsenal Tank Plant in 1996, General Dynamics relocated parts of its tank maintenance operation to the Lima Army Tank Plant.

The U.S Marine Corps expeditionary fighting vehicles are also manufactured in Lima. For this reason, the U.S. Defense Department changed the plant's name to the Joint Systems Manufacturing Center in June 2004. Currently, the modified M1A1 Abrams (1986–present) is being produced for the U.S. Army and other NATO forces.

LIMA LOCOMOTIVE WORKS

Beginning in Lima, Ohio, in 1877, the Lima Machine Works started manufacturing agricultural harvesting and sawmill equipment. The next year, the company received a contract to fabricate a steam locomotive modeled after a design by railroad engineer Ephraim Shay (1839–1916). The Shay locomotive was engineered to be geared down to provide more slow-moving pulling power. This proved to be an essential requirement for transporting logs and lumber.

The first Shay locomotive was completed by Lima Machine in 1880. After testing, it proved to be exactly what the lumber industry needed. Lacking the resources to meet the immediate new demand for his locomotive, Ephraim Shay granted the Lima Machine Works a license to manufacture the locomotive, which was now referred to simply as a "Shay." Lima Machine proceeded to expand its foundry, fabrication and assembly facilities to ramp up production. As a result, the Shay locomotives were now able to be delivered as needed to the ever-expanding American lumber industry. Within two years, Shay steam locomotives became Lima Machine Works' number one manufactured item.

Over the next decade, the company turned out more than 300 locomotives, which were then shipped throughout the United States and Canada. In 1892, Lima Machine Works was reorganized and renamed the Lima Locomotive & Machine Company. By 1893, it had successfully manufactured 450 locomotives.

Industrialist Joel Coffin bought the company in 1916 and renamed it the Lima Locomotive Works. In addition to the Shay locomotive, Lima Locomotive Works is best remembered for manufacturing the advanced steam locomotive known as "Super Power." Designed by mechanical engineer William E. Woodard (1873–1942), it went into production

Opposite, top: Lima Locomotive Works locomotive assembly building, circa 1922. *State Library of Ohio.*

Opposite, bottom: Berkshires locomotives, like this one seen about 1949, were regarded as a Super-Power class of steam engines. *State Library of Ohio.*

Above: After World War II, Lima-Hamilton began manufacturing diesel locomotives. This locomotive is shown circa 1947. *State Library of Ohio.*

in the 1920s. The Boston & Albany Railroad, a subsidiary of New York Central, bought the first forty-five Super Power locomotives, consisting of a 2-8-4-wheel arrangement and known for their pulling ability. These engines were known as the "Berkshire" on most railways. More than 615 Berkshire locomotives were built and placed into service by 1949.

During World War II, Lima Locomotive Works began manufacturing M4A1 Sherman tanks in 1942 for use by our British allies. After the war, Lima Locomotive merged with the General Machinery Corporation of Hamilton, Ohio, and on July 30, 1947, the new company emerged as the Lima-Hamilton Corporation. On May 13, 1949, the last steam locomotive came off the line, and Lima-Hamilton turned its focus to manufacturing diesel locomotives. The Lima plant continued manufacturing heavy construction equipment and cranes.

The Baldwin Locomotive Works of Philadelphia and Lima-Hamilton were merged on September 11, 1951, to form the Baldwin-Lima-Hamilton Corporation. Lima-Hamilton completed its last diesel engine the following day. But in May 1956, Baldwin-Lima-Hamilton completely bowed out of the locomotive manufacturing business.

GE AVIATION

GE Aviation, a manufacturer of jet engines, has its main offices in Evendale, Ohio, near Cincinnati. GE Aviation is a subsidiary of General Electric and is one of the top-ranked suppliers of jet engines, primarily for use with commercial aircraft. Most people think of lights and motors in relation to General Electric. But it also has an extensive history in the development of turbine engines.

In 1903, aviation engineer Sanford Alexander Moss (1872–1946) started working for GE's steam turbine department and was instrumental in the development of turbo superchargers for aircraft. Moss also developed a high-RPM supercharger that was propelled by the flow of the engine exhaust. This supercharger was successfully tested at Wright Field in Dayton, Ohio, in 1918, and GE was awarded its first military contract to supply superchargers.

After World War I, GE developed an improved supercharger that captured energy from the rotating piston crank shaft and funneled it to compress the air in the cylinders of the engine. This resulted in the engine burning more fuel, thus delivering more power. By the start of World War II in 1939, superchargers were regarded as an essential requirement for advanced aircraft. This placed GE in a position to be a prime manufacturer and supplier of this technology.

In 1939, Air Commodore Sir Frank Whittle (1907–1996), an English engineer, inventor and Royal Air Force (RAF) officer, developed the first turbojet engine. This was an entirely new kind of gas turbine. The engine uses a series of different-size fan blades to compress intake air into a combustion chamber. By detonating a high-octane fuel in this chamber, tremendous pressure is released. As these gases exit the tailpipe, an enormous amount of thrust is created. Compared to a conventional supercharged, piston-powered aircraft, planes with Whittle's jet engine were easily able to be propelled at twice the speed. GE proved to be a near perfect industrial partner to develop jet engines after Whittle's model W.1 jet engine was demonstrated to the U.S. Army Air Corps in 1941. After a manufacturing license was issued, several W.1 test engines were shipped to the United States for further study and refinement. GE quickly commenced production of improved models to meet American standards. While only a small number of jet engines were produced, a more powerful I-40 model followed in 1943. It was used to power the first U.S. combat-capable jet fighter, the Lockheed P-80 Shooting Star, which saw limited service before the war ended.

Top: Ohio-based GE Aviation is a manufacturer of jet engines used in both military and commercial aircraft. A jet engine for NASA is seen here about 2019. *National Museum of the U.S. Air Force.*

Bottom: This U.S. Air Force transport is propelled by GE Aviation jet engines built in Evendale, Ohio. Photograph circa 2010. *National Museum of the U.S. Air Force.*

Following World War II, GE turned its attention from supercharger technology and began focusing its research on further development of the high-performance, exhaust-driven turbo system (jet engine). In 1964, Curtiss-Wright, Pratt & Whitney and GE submitted their design for an engine to power the Lockheed C-5 Galaxy, a large-body, high-wing military cargo transport aircraft. In 1965, the GE engine model F-103 was selected, and this resulted in a civilian version known as the model CF6, which powers the McDonnell Douglas DC-10 aircraft. Lockheed later switched to using the Rolls-Royce RB211, but the DC-10 maintained use of the CF6. This success was the springboard to extensive sales and installations on a wide range of large aircraft, including the Boeing line of 747s.

GE's model CF6 jet engine class remains commercial aviation's longest-running jet engine program, with service beginning in 1971. Model CF6 was GE's first successful power plant for use on commercial widebody aircraft. The current fleet of the U.S. presidential *Air Force One* 747s is powered by CF6. The GE model CF6 jet engine class powers approximately 70 percent of the world's widebody aircraft.

The CF6 is the leading engine of choice for new and conversion widebody freighters due to its proven durability, reliability and performance. As older passenger aircraft are converted into freighters, the CF6 engine offers significant advantages, including the highest widebody engine departure reliability rates, fewer flight disruptions and extensive used material availability for lower maintenance costs.

The U.S. presidential *Air Force One* 747 fleet is powered by GE model CF6 jet engines. The plane is seen here about 1995. *National Museum of the U.S. Air Force.*

GE Aviation is one of America's top exporters, supplying engines for a wide range of commercial and military aircraft, boats and hydrofoils. Its jet engines can also be found in helicopters, hovercraft, speedboats and industrial power generators. GE also developed a J85 series that is a popular power plant for private jets. History will record that the key to GE Aviation's success in the industry can be understood by looking at its commitment and heavy investment in research and development and its maintenance of stringent quality-control standards.

9

BUGGIES AND BIKES YIELD TO CARS AND TRUCKS

Columbus Buggy

During the nineteenth century, the Columbus Buggy Company was recognized as the country's premier manufacturer of passenger buggies. In 1870, the Iron Buggy Company was established by three business partners, Clinton D. Firestone and brothers George and Oscar Peters. Their mission was to build and sell inexpensive buggies to the working class. Within a short time, their business plan was met with immediate success. The favorable results were primarily due to a design engineered by the Peters brothers. During the first years of operation, they managed to sell 237 buggies. After a fire destroyed the buggy shop in 1874, the three partners decided to construct another fabrication shop. But in 1875, they sold the Iron Buggy business to another local buggy manufacturer, the Buckeye Buggy Company.

In 1875, armed with $20,000 in capital, the three partners proceeded to establish the Columbus Buggy Company and Peters Dash Company. These businesses specialized in making buggies that were affordable and in various price ranges. Buggy sales began to rapidly increase, and by mid-1878, the company had a staff of 255 employees turning out one hundred buggies weekly.

Within eight years, the company was employing a staff of more than one thousand people. The production line was turning out more than 25,500 buggies annually. In 1883, the Canton-based Timken Company

Clinton D. Firestone was one of three business partners who founded the Columbus Buggy Company. He is seen here circa 1890. *Columbus Metropolitan Library.*

filed an infringement lawsuit against Columbus Buggy. The lawsuit initially sought liquid damages and focused on the allegation that the company had stolen a patented design for an "automatic" spring part used for suspension in its carriages. The Columbus Buggy Company responded by filing a countersuit, but the judge decided in the plaintiff's favor in December 1884. The issue was settled, and the Columbus Buggy Company paid damages and for the model year 1886 began using Timken spring suspension components.

By 1888, the company occupied the entire city block between High Street and Wall Street and, within the buggy industry, was regarded as the largest manufacturer of buggies worldwide. At this time, the marketing department started heavily advertising its vehicles in magazines and newspapers with national circulations.

The company's golden year was 1892, when it employed a staff of 1,200 people. At this time, its peak production was one hundred buggy vehicles each day. The buggies were being exported to other countries. The company exhibited at the 1893 World Columbian Exhibition in Chicago, where it made it known that there were Columbus Buggy Company dealerships in every state and in a number of countries. But in light of the country's financial panic in 1893, the company's buggies came to be regarded as expensive and overpriced.

By 1900, there was serious competition in Columbus proper, because there were an additional two dozen other buggy manufacturers in town. In the late nineteenth century, Columbus was for buggies what Detroit was for automobiles in the twentieth century. With the buggy industry in Columbus operating three shifts, it was estimated that 17 percent of all buggy vehicles in service around the world were manufactured in Columbus. The Columbus Buggy Company was one of the largest businesses in the city and was in need of more space. For this reason, and with his interest starting to focus on manufacturing automobiles, Clinton Firestone relocated his production to a new facility at 400 Dublin Avenue.

The introduction of the automobile placed the Columbus Buggy Company at a strategic disadvantage compared to manufacturers that had access to one of the Great Lakes. Companies in cities like Chicago, Cleveland and Detroit

A Columbus Buggy Company luxury model buggy is seen here about 1890. In the late nineteenth century, Columbus was for buggy transportation what Detroit became for automobiles in the twentieth century. *Library of Congress.*

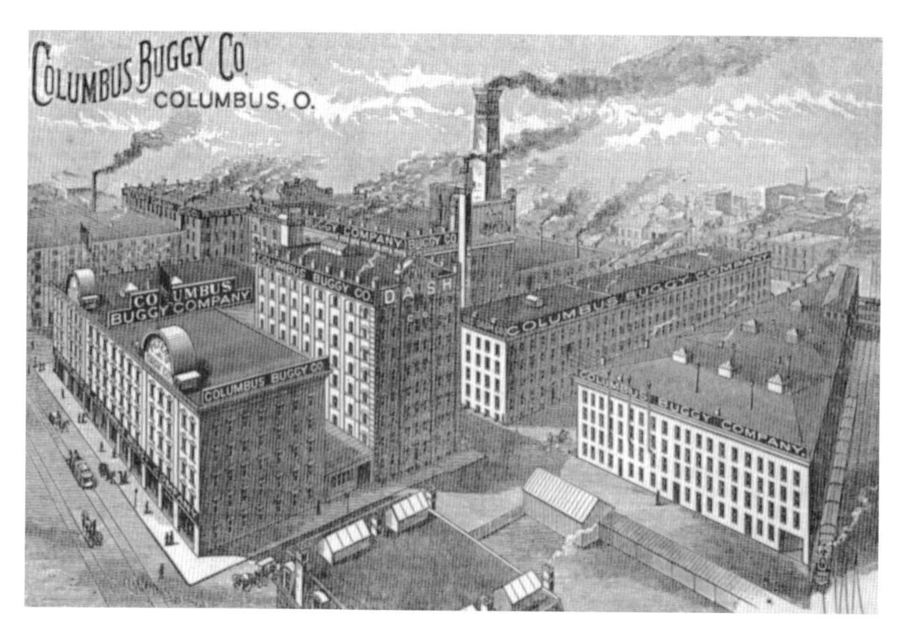

The Columbus Buggy Company manufacturing complex on Dublin Avenue (now Nationwide Boulevard), circa 1889. *Columbus Metropolitan Library.*

MODEL 1220

The Special Features and Refinements

OF THE 1912

COLUMBUS ELECTRIC

"The Car Supreme"

Cannot possibly be enumerated fully in this space. Some of the more important ones are as follows:

Large, roomy body. Luxurious trimmings and deep comfortable cushions. Two dome lights operated by opening and closing right hand door. Automatic electric heater. Handsome cut flower vase. Reliable clock. Neat, serviceable toilet case. Adjustable mirror, enabling driver to see traffic approaching from the rear. Stationary front seat, semi-divided pattern. Perfect spring suspension. "I"-beam front axle, with ball-bearing steering knuckles. Wheel base 92 inches with 56 inch track. Tires 34 x 4 inches, "nobby tread" on rear wheels.

Write for Catalog 61-E

The Columbus Buggy Company

561 Dublin Avenue, Columbus, Ohio

Columbus Buggy Company had an excellent reputation for manufacturing quality electric vehicles. Photograph circa 1912. *Columbus Metropolitan Library.*

had better access to the resources required to manufacture automobiles, such as steel and rubber, brought in by ships operating on the Great Lakes.

Following the turn of the century, electric automobiles were regarded as the new mode of transportation and were gaining in popularity. So, in 1903, the 400 Dublin Avenue factory began manufacturing the company's first electric vehicle, the 1903 Columbus Electric Model No. 1000 Coupe. The year 1905 marked the beginning of a new era that included several other models of electric vehicles the company was turning out on a larger scale. The company's advertisements targeted women by pointing out how simple, safe and easy the vehicles were to operate and that they were quiet. But electric vehicles had three main disadvantages: they could only reach a top speed of twenty miles per hour, they had a limited range and the batteries required several hours to be recharged.

Other vehicles manufactured included the No. 3003 Firestone Motor Buggy. This lightweight buggy was marketed to physicians as a medical phaeton carriage. The Columbus Station Wagon No. 1102 was to be driven by a hired chauffeur while the passengers rode in a comfortable, enclosed compartment.

The Columbus Buggy Company was looking to quickly gain experience in the engineering and manufacture of cars. So, in 1907, it acquired the Springfield Automotive Company, located in Springfield, Ohio. This automaker had previously manufactured the gasoline-powered Bramwell automobile. In 1909, the company hired race-car driver Lee Frayer (1874–1938), whose job it was to design a full-sized car. Frayer's assistant was eighteen-year-old Columbus native Eddie Rickenbacker (1890–1973).

A model that did fairly well in the market was the Firestone-Columbus. This gasoline-powered car was designed with families in mind. Production of the Firestone-Columbus began in 1909, with nearly five hundred vehicles sold in the first year after its introduction. It is believed that the Columbus-Firestone was the first car to have the steering wheel mounted on the car's left side.

The future of Columbus Buggy was looking promising, then the local Scioto River overflowed during the 1913 flood that devastated the lower-elevation areas of the city's west side. The flood disrupted production so much that the company was forced into bankruptcy and closed its doors.

But in 1914, the same year that Clinton D. Firestone died, a new establishment, the New Columbus Buggy Company, was incorporated with financing from a new group of investors. The corporation was under the leadership of the former company's creditors and continued manufacturing

Columbus native Eddie Rickenbacker, seen here about 1905, assisted car designer C.C. Bramwell in developing a gasoline engine passenger car for the Columbus Buggy Company. *Columbus Metropolitan Library.*

at the Dublin Road plant. The Firestone Electric and the Columbus-Firestone production lines continued well into 1915, when manufacturing activities were halted and the company discontinued operations. The Columbus Buggy Company had a significant influence on early methods of automobile manufacturing. In addition to Eddie Rickenbacker, another notable employee who got his start in automobile manufacturing was Harvey S. Firestone, whose work and career were detailed earlier in this book.

HUFFY CORPORATION

Huffy Corporation dates to 1887, when George P. Huffman (1862–1897) acquired the Davis Sewing Machine Company in Watertown, New York. The company manufactured sewing machines until 1890, when Huffman relocated the sewing machine factory to Dayton, Ohio, Huffman's hometown. In 1892, the Davis Sewing Machine Company produced its first "Dayton" brand bicycle.

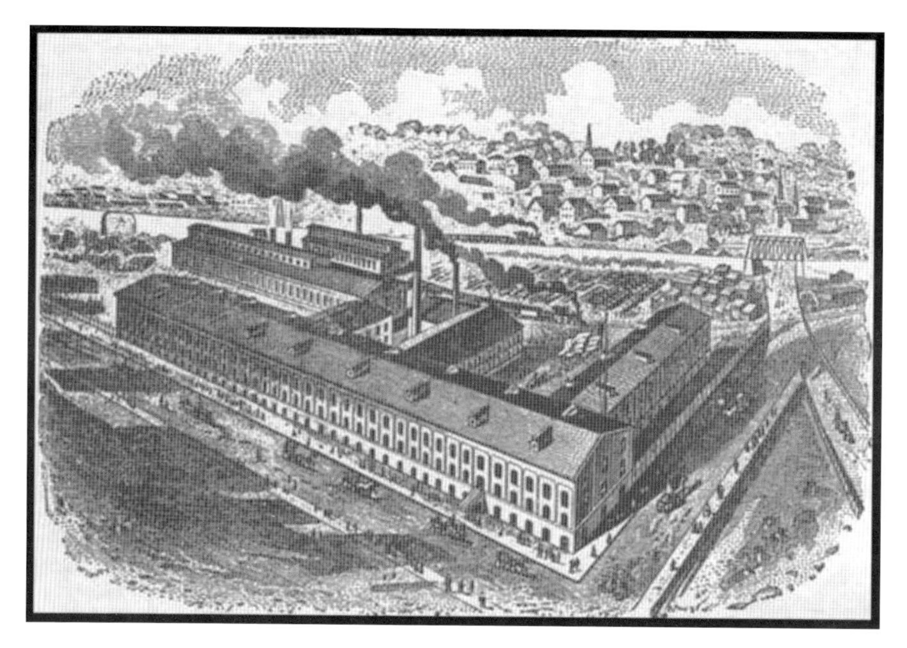

George Huffman's Davis Sewing Machine Company was a major Dayton, Ohio employer. This illustration shows the factory about 1905. *Dayton and Montgomery County Library.*

In 1924, George's son Horace Sr. (1885–1945) established the Huffman Manufacturing Company to manufacture steel bicycle rims and entry-level bicycles. The facility also served as a sales outlet for Davis bicycle parts. Horace Sr. learned the manufacturing business while working for his father's Davis Sewing Machine Company, which was sold in 1925. Initially, Huffman Manufacturing produced and sold bicycles using the Dayton brand name and continued this model line until 1949. In the mid-1920s, the company started manufacturing bicycles with the Huffman brand name.

Huffman also manufactured a line of automobile service station equipment. The company leadership incorporated in 1928. During the Great Depression, Horace Huffman unveiled a strategic plan to manufacture bicycles, which was an affordable means of transportation in the prewar years. Although the production rate grew from a dozen bicycles to two hundred bicycles a day within two years, it lagged the rate of competitors. Realizing the true nature of the problem, Horace M. "Huff" Huffman Jr. (1914–1996) transformed the production process to a straight-line conveyor assembly line. Within two years, bicycle production doubled. The increased production brought new business, including Western Auto

A Davis Sewing Machine Company "Dayton model 28" bicycle in 1922. *Dayton and Montgomery County Library.*

During World War II, Huffman Manufacturing Company produced four thousand bicycles for the U.S. War Department. The bicycles saw service all over the world. Photograph circa 1943. *Dayton and Montgomery County Library.*

Company, and returned lost business, including the Firestone Tire and Rubber Company.

At the start of World War II, production priorities shifted to components for artillery shells. In 1942, the War Department ordered four thousand Huffman bicycles. During that time, much of the manufacturing labor was provided by women while men were serving in the armed forces.

Company founder Horace Huffman Sr. died in Dayton in 1945, and Horace Huffman Jr. assumed his father's position as corporate president. The 1949 postwar recession hit companies hard; Huffman survived the downturn with key product developments. Huffman introduced the Huffy Convertible bicycle, which quickly gained popularity nationwide and made Huffy an icon in the bicycle industry. That marked the first of many notable bicycle innovations. By 1960, Huffman was the third-largest bike manufacturer in America. The second development introduced by Huffman was the manufacture of lawnmowers.

In 1962, Horace Huffman Jr. was named chairman of the board, and in 1964, the Huffman corporate offices were moved to Miamisburg, Ohio, a suburb south of Dayton. By 1973, the company was employing more than 2,550 workers. During the 1960s and well into the 1970s, adults enjoyed bicycles for recreation, exercise and easy, cost-effective transportation. Prior to 1970, over half of the bicycles in America were purchased from small, independent bicycle shops that offered personal service and maintenance. In the 1970s, mass merchandising retail discount chains opened, bringing a new market for selling bicycles. Huffman focused its efforts on this new market, engineering an affordable ten-speed bicycle with easy assembly and easy maintenance.

Following a difficult period during the recession of the mid-1970s, Huffy rebounded with substantial growth in the 1980s. Bicycle manufacturing operations in Celina included advanced robotics and assembly line equipment. These modifications resulted in increased production of an additional five thousand bicycles a day while reducing production costs by 14 percent. At its peak, the bicycle division manufactured over two million bicycles annually, making it the world's largest bicycle manufacturer.

But by 1995, bicycle business was just one portion of the diversified Huffy Corporation. The overall bicycle market shifted significantly, with independent bicycle shops holding a 25 percent share of the market and mass retailers holding 75 percent. Many trends started at bike shops, including mountain bikes. Companies like Huffy brought these high-end trends to the general public through mass retailers at lower price points.

If you were lucky enough to own a Huffy Radiobike, you had the number-one bike of atomic age. Photograph circa 1955. *Dayton and Montgomery County Library*.

The United States did not place tariffs on low-cost Chinese imports during this period, causing significant pricing pressures on American-made bicycles. One by one, brands moved manufacturing offshore. Huffy was the last of the "big" brands to move manufacturing from the United States. While production continues offshore, other Huffy business operations, including all of the product design and development, remain in Dayton, where it began.

Alexander Winton

Alexander Winton (1860–1932) was a bicycle, automobile and diesel engine designer and inventor who has been mostly forgotten. At the age of twenty-one, he emigrated from Scotland and settled in Cleveland, Ohio, and established the Winton Bicycle Company. His interest turned to designing an automobile, and in 1896, he built his first motorized vehicle. He organized the Winton Motor Carriage Company in 1897 and the next year made the first commercial automobile sale in America.

Winton invented the semi-truck in 1898 and sold his first manufactured semi-truck the following year. When he initially began manufacturing cars,

Left: Alexander Winton, shown here about 1895, was an innovative bicycle and automobile designer and inventor. *Cuyahoga County Public Library*.

Below: Winton in one of his race cars, used to help promote and test vehicles innovations, circa 1900. *Cuyahoga County Public Library*.

Opposite: A Winton luxury touring car, circa 1912. *Cuyahoga County Public Library*.

he wanted to ship directly to customers in order to not place any miles on the vehicles. He not only developed a car hauler for his company's use, but within a short time he was also selling these car haulers for use by other car manufacturers.

Winton participated in cross-country auto tours and races to promote and test the innovations incorporated in the design of his automobiles, and he registered over one hundred patents. America's first cross-country automobile tour was completed in 1903 using a Winton motorcar. By that year, the company was employing a staff of over 1,200 workers. Winton Motor Carriage Company built and sold twenty-two automobiles in its first year in business and increased production to over one hundred vehicles the following year.

Henry Ford became a major name in automobile design and manufacturing, thanks to Winton granting Ford use of his steering wheel design prior to a race in 1901. Despite Winton's reputation as an automotive innovator, each Winton vehicle was custom-made. In contrast, automotive assembly lines made vehicles less expensive and the industry more competitive in the 1920s. In 1924, Winton completely withdrew from manufacturing automobiles.

Although no more Winton vehicles were built after 1924, Winton himself continued to be an innovator in the development of both diesel and gasoline engines. He founded the Winton Engine Corporation; in 1930, it became a subsidiary of the General Motors Corporation.

THE FISHER BODY COMPANY

The Fisher Body Company was established on July 23, 1908, by brothers Fred J. Fisher (1878–1941) and Charles T. Fisher (1880–1963) and their uncle Albert Fisher (1864–1942). All three men were natives of northern Ohio, but they established their auto-body-building company in Detroit, Michigan, because their major customers were Cadillac, Ford Motor Co., Buick, Studebaker, Packard and other car companies. In the early days, a car body was simply a modified horse-drawn coach. Fisher bodies were sturdy and entirely enclosed to protect passengers from the elements, and this made cars more marketable to families.

The company experienced immediate success and was producing over one hundred thousand auto bodies by 1913. After the departure of Albert, the five remaining younger Fisher brothers—William, Lawrence, Edward, Alfred and Howard—were brought into the company. Each brother brought a different set of skills but shared a common goal: high-quality craftsmanship. By 1914, the Fisher Body Company was the world's largest auto-body

The Fisher Body Company's Cleveland, Ohio plant in 1944 produced wing components for the B-29 Superfortress bombers during World War II. *Library of Congress.*

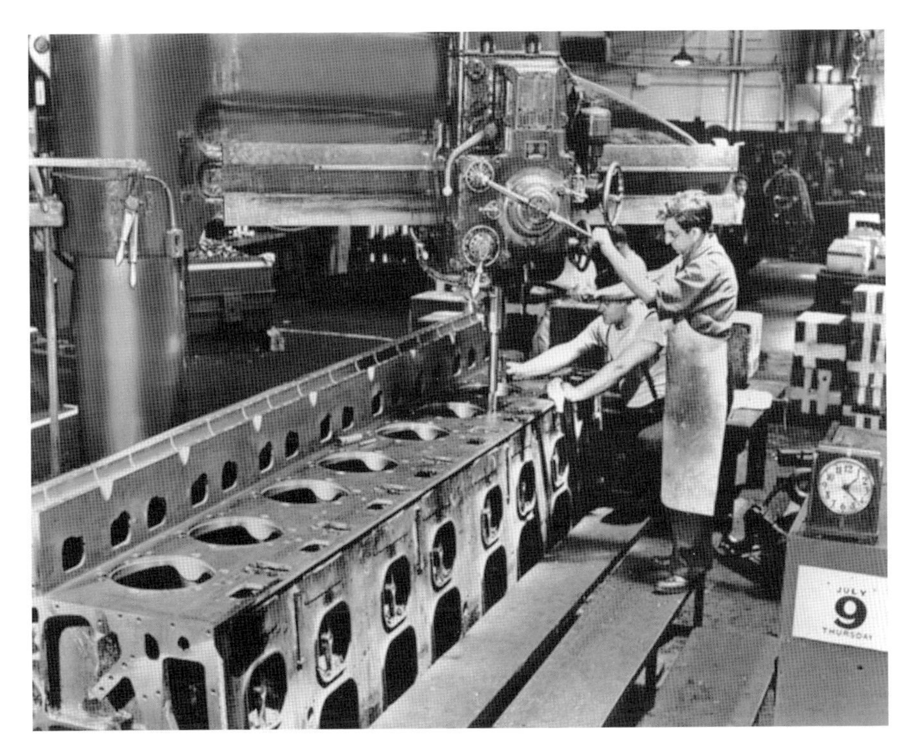

Workers at the Fisher Body Company's Cleveland plant in 1943 are machining large-scale diesel engines used to power submarines during World War II. *Library of Congress.*

producer in the world. At one point, the company payroll included well over one hundred thousand office, skilled and unskilled workers. Fisher Body manufactured over two thousand airplanes after America entered World War I. By 1918, the company's output had increased to nearly three hundred thousand car bodies. Also, by 1918, the firm was manufacturing bodies for most of the major automakers.

In 1919, General Motors president William C. Durant acquired a three-fifths interest in the Fisher Body Company and became the sole owner in 1926. Fisher Body was now the in-house coach-fabricating division of GM. The Fisher Body Division Company expanded its facilities into several other states, including Columbus and Cleveland, Ohio.

With the outbreak of World War II, the Fisher Company began to produce military ordnance for the American war effort. By the end of the war, the company had manufactured 1,200 Pershing tanks, 11,358 Sherman tanks and 5,368 M10 tank destroyers. The firm also produced B-29 Superfortress bombers at a new facility in Cleveland. This was the model of bomber used to deliver and drop the atomic bombs on Japan that completely ended World War II in the Pacific theater.

The Fisher Company continued to profit in the postwar years and into the 1980s. It was instrumental in manufacturing the first airbag at plants in both Detroit and Euclid, Ohio. In addition to automatic seatbelts, the Fisher Body Division developed sunroofs for GM automobiles and at one point operated seven manufacturing facilities throughout Ohio.

General Motors continued to use the "Body by Fisher" logo, and the name could still be seen on GM cars into the mid-1990s. In 1984, the Fisher Body Division was dissolved into other GM divisions.

WILLYS-OVERLAND AND JEEP

Industrialist John North Willys was the founder of Willys Overland Motor Company headquartered in Toledo, Ohio. He is seen here circa 1910. *University of Toledo Library.*

In 1908, John North Willys (1873–1935) acquired the failing Indianapolis-based Overland Automotive Division of Standard Wheel Company. The next year, Willys purchased the Marion Motor Car Company of Indianapolis and combined the assets to manufacture and sell four thousand cars. Willys's keen business sense and ingenuity had taken a nearly bankrupt company and returned it to a profitable enterprise by the next fiscal year.

With his focus on further increasing manufacturing productivity, Willys invested company profits in 1909 to acquire the bankrupt Pope Motor Car Co. of Toledo, Ohio. At that time, the Pope auto plant was the world's largest car-manufacturing building. John Willys relocated the entire operation to Toledo and was now able to substantially increase car production. In 1912, the company was renamed the Willys-Overland Motor Company.

Willys-Overland's car production had exceeded 141,000 units by 1915. Willys-Overland's automobile sales were second only to the Ford Motor Company for five consecutive years starting in 1912. In 1917, Willys established the Willys Corporation as a separate holding company to facilitate continued expansion and enable the company to manufacture approximately 1,000 cars daily by the end of the year. At this time, the

A Willys-Overland steel forge press operator fabricating vehicle components, circa 1920. *University of Toledo Library.*

company had manufacturing plants in Ohio, New York and Michigan and employed thirty-eight thousand workers.

John Willys expanded his holdings by purchasing the Curtiss Airplane and Motor Corp. of Buffalo, New York, and enlarged the facility for wartime production. Willys-Overland was a major supplier of trucks, airplanes and airplane engines during World War I.

After the war, the Willys Corporation was grossly overextended because of its rapid expansion program. In 1920, the company was losing money and was in serious debt. As a result, Willys-Overland was forced into receivership, and the creditors selected New York banker Walter P. Chrysler as vice-president and general manager. Chrysler liquidated a variety of major assets and completely reorganized the company. Always a good salesman, Willys crisscrossed the country recruiting new dealers and motivating existing dealers. The efforts of both men proved that the company was profitable once again.

Willys was soon back in complete control of the company following Chrysler's departure to establish an automobile company under his own name. Willys-Overland's success continued until 1929, when it manufactured well over 315,000 vehicles. That year, just prior to the October stock market crash, Willys retired from the company but retained his position as chairman of the board and sold his common stock shares for $25 million (equivalent to $412 million in 2023). He resigned as president but remained chairman of the board. In 1935, Willys died following a heart attack while living in New York City. A year after John Willys's death and coming out of bankruptcy, the company was reorganized as Willys-Overland Motors Inc.

At the outbreak of World War II, Willys-Overland Motors received a federal contract to manufacture military jeeps for the war effort. The jeep was a lightweight, four-wheel-drive vehicle originally based on a design by the Bantam Car Company. By the end of the war in 1945, Willys had turned out over 360,000 vehicles.

Following the war, the Willys Jeep Wagon went into production for civilian use. From 1946 until 1965, 300,000 Jeep wagons and 200,000 Jeep trucks came off the assembly line. Industrialist Henry J. Kaiser acquired the company in 1953, and the name was changed to Willys Motor Company. For the next ten years, Jeep wagons and trucks were produced with the Willys trade name. In 1963, the company was renamed Kaiser-Jeep Corporation.

In 1970, American Motors Corporation acquired control of Kaiser-Jeep, and Chrysler Corporation acquired American Motors in 1987. Eleven years later, European carmaker Daimler-Benz merged with Chrysler Corporation, and the world's fifth-ranked car manufacturer, Daimler-Chrysler, was established.

Daimler-Chrysler opened a new state-of-the-art manufacturing facility in Toledo in 2001. Among the new models of Jeep that have rolled off this assembly line include the Jeep Commander, Compass, Patriot, Wrangler and Gladiator. In 2007, Cerberus Capital acquired an 80 percent share of the American Chrysler division of Daimler-Chrysler. The Italian automaker

These women are doing the front-line manufacturing work on the military jeep assembly line in 1942 at the Willy-Overland plant in Toledo, Ohio. *Library of Congress.*

Fiat acquired total ownership of Chrysler in 2014, and the Jeep brand was manufactured by Fiat Chrysler Automobiles. As of January 2021, Fiat Chrysler, USA, is now a subsidiary of the multinational carmaker Stellantis, headquartered in Amsterdam, Netherlands. But its Jeep brand automobile is still made in Ohio at the Toledo Assembly Complex, which operates two shifts, uses 616 assembly robots and employs a staff of 6,290 workers.

FIRE ENGINES COME OF AGE IN OHIO

For hundreds of years, fighting fires was done manually and with minimal success, until three men in Cincinnati, Ohio, seriously addressed the problem. The combined skill of Alexander B. Latta, a locomotive builder; Abel Shawk, a locksmith; and Miles Greenwood, who owned a small iron foundry, produced a firefighting steam engine that delivered water pressure in ten minutes. Following a demonstration, the Cincinnati City Council voted to approve purchasing a firefighting steam engine.

On January 1, 1853, the Cincinnati Fire Department took delivery of the world's first practical steam fire engine. Instead of relying on firefighting volunteers, on April 1, 1853, Cincinnati organized the first compensated professional fire department in America. Often overlooked is Ohio's heritage as an innovator and manufacturer of fire engines for municipal fire departments across the United States.

The Ahrens-Fox Fire Engine Company was a Cincinnati fire truck manufacturer established by partners John Ahrens and Charles Fox, who fabricated their first motorized fire engine in 1911. Ahrens-Fox fire engines were easily recognized because of the air-filled, chromed sphere positioned over the pump that significantly reduced outlet pressure fluctuations from the piston pump. Ahrens-Fox is now a part of Michigan-based HME Inc.

The Seagrave Corporation manufactured firefighting apparatus from 1891 until 1965 in Columbus, Ohio, used by fire departments throughout the United States. After World War II, Seagrave introduced a three-man cab that alleviated firefighters having to travel to the site of a fire standing on running boards mounted on the side of the truck. Seagrave was moved to Wisconsin in 1965.

SUTPHEN

Another great Ohio name well known to the nation's firefighting industry is Sutphen. Founded in 1890 in Columbus by C.H. Sutphen, Sutphen Corporation currently stands as the oldest family-owned and -operated fire apparatus manufacturer in the United States. Built on the foundations of family, innovation and service to firefighters and communities around the world, Sutphen Corporation has been an industry staple since 1890.

The family-owned business has never been reorganized and has not been owned or operated by anyone other than a Sutphen family member. Today, Sutphen is still based in central Ohio and led by family members in nearly every aspect of the business, including the following:

Drew Sutphen, president
Julie Sutphen Phelps, vice-president
Judi Sutphen, contract administration

C.H. Sutphen (*standing, far left*), the founder of Sutphen Corporation, pictured with a steam fire pumper and crew, circa 1898. *Courtesy Sutphen Corporation.*

Shelby Sutphen Chambers, strategy and culture advancer
Dan Sutphen, sales
Harry Sutphen, sales
Dareth Sutphen Fowler, service and technical division
Steven Phelps, accounting
Andy Herb, sales
Scott Herb, sales
Jim Holland, engineering

Prior to Drew and Julie's leadership, the business was run by their fathers, grandfathers and great-grandfathers, as well as numerous other family members. Sutphen, much like the firefighting industry, is rooted in family and steeped in tradition. One of the most influential members of the Sutphen family was Tom Sutphen, Julie's father and Drew's uncle. Tom was a true innovator and left his legacy on the industry by inventing the mid-mount aerial apparatus. A first of its kind, this apparatus revolutionized the fire industry. Sutphen works to continue his legacy by manufacturing the most sought-after mid-mount aerial on the market.

Today, Sutphen operates out of six manufacturing facilities, five of which are in central Ohio. Sutphen's Dublin, Ohio facility manufactures aerial platforms and aluminum custom pumpers while also housing corporate

Top: A current Sutphen model midmount aerial platform at a fire scene with the Columbus Division of Fire. *Courtesy Sutphen Corporation.*

Bottom: The Sutphen SPH 112 is a midmount aerial platform custom built for NASA's Kennedy Space Center in Florida. *Courtesy Sutphen Corporation.*

offices. In Hilliard, Ohio, Sutphen manufactures aerial platforms as well as its aerial ladders and stainless-steel pumpers. Urbana, Ohio, has three Sutphen facilities, including its Urbana Pumper Division, Sutphen Chassis Division and Sutphen Service and Technical Division.

Sutphen also operates a facility in Lake Ariel, Pennsylvania, focused on custom pumpers, commercial chassis apparatus, tanker apparatus, rescue apparatus and East Coast customers. In the fall of 2021, Sutphen broke ground on a new Urbana facility that will combine its three Urbana operations into a single 185,000-square-foot facility. The new venue is slated to open in the summer of 2023 and will expand production capacity while bringing the three Urbana facilities under one roof.

Sutphen Corporation is a sole-source manufacturer, meaning that the company builds each fire truck from the ground up. From its own extreme-duty chassis to its aerial ladders and everything in between, Sutphen provides its customers a robust and heavy-duty product.

Sutphen sells its fire apparatus to departments across the country and around the world through a widespread dealer network. Similar to a local car dealership, Sutphen has dealerships across the country and around the world that assist customers in creating truck specifications and working through the purchasing process.

As a family-owned company, Sutphen prides itself on the entire Sutphen experience, which includes a family experience, a quality product and service after the sale.

HONDA OHIO

Honda is an extremely dominant but quiet fact of life in the state of Ohio. Honda started with a motorcycle manufacturing plant in Marysville and now has three auto plants in the state. In addition, Honda manufactures engines and transmissions in Ohio at the company's Anna Engine Plant and Honda Transmission Manufacturing of America in Russells Point. In 2020, a Platinum White Pearl Honda Accord Hybrid rolled off the production line at the Marysville Auto Plant, symbolizing the twenty-millionth automobile manufactured in Ohio.

But the real story of Honda in Ohio begins with two men: former mayor of Columbus and former four-time governor of Ohio, James Allen Rhodes (1909–2001); and Soichiro Honda (1906–1991), the founder and president of Honda. Governor Rhodes had long acquired a deserved reputation for being known as a "smokestack chaser," which meant he saw jobs and industry as the way to maintain a vibrant state economy. Soichiro Honda had as his philosophy that he wanted to manufacture automobiles in the markets where they were being sold.

As a result, in 1976, at a Wendy's restaurant in Columbus, Governor Rhodes sat down with Mr. Honda and his staff for burgers at 10:30 a.m.

Left: Governor James Allen Rhodes (*right*) and Soichiro Honda, the founder and president of Honda, in 1976 saw a mutual benefit to Honda manufacturing in Ohio. *Columbus Dispatch Newspaper*.

Below: The Honda Marysville, Ohio plant assembly line producing model year 1982 Honda Accords. *State Library of Ohio*.

for an early lunch. Rhodes wanted Honda to build its first American plant in Ohio. Rhodes then escorted Honda and staff to a site just outside a small rural town in Union County. There, Honda was able to inspect a flat acreage area that quickly impressed upon him that he could build a manufacturing plant on the site. Although the two men were an unlikely pair, Governor Rhodes and Mr. Honda forged a bond that resulted in Honda of America purchasing an initial 217-acre property in 1977 located just outside Marysville for the purpose of constructing a motorcycle manufacturing plant.

Automobile manufacturing began in Marysville, Ohio, on November 1, 1982, with the assembly of the Honda Accord. This marked the first Japanese automobile produced in America. Currently, the Marysville Auto Plant has two production lines and is Honda's largest vehicle manufacturing plant in America. In 1989, Honda added its second Ohio auto manufacturing plant, known as the East Liberty Auto Plant. It produces the Honda CR-V and the Acura RDX and MDX. Honda started Ohio production of the Acura NSX supercar in 2016 at the Performance Manufacturing Center, also located in Marysville.

Honda's initial $35 million investment (equivalent to $140 million in 2023) to manufacture motorcycles in Marysville in 1979 has grown to over $13 billion in its statewide operations. James Rhodes's objective has significantly exceeded his expectations, with over fifteen thousand people employed as Honda associates across the state. By comparison, General Motors has a payroll of seventeen thousand hourly employees across all of Michigan. Honda employees repeatedly demonstrated their approval of the hourly compensation, benefits and working conditions. They have refused to adopt a union so many times that the UAW no longer maintains an office in Marysville.

Hercules Motors

Hercules Motor Manufacturing Company was established in 1915 in Canton, Ohio, to manufacture lightweight, high-speed gasoline engines for the truck industry. Their engines were soon in high demand and became standard equipment for many major truck manufacturers. That demand also spilled into farm machinery, heavy construction equipment and oil-field equipment and into other industries because of the engine's reputation for

Hercules Motors Corporation manufactured gasoline and diesel equipment engines with a reputation for reliability. Photograph circa 1922. *Cuyahoga County Public Library.*

dependability and the company's ability to adapt its engines to just about any need or application.

In 1922, the company reincorporated and restructured its general operations, and the result was the Hercules Motors Corporation. During the time between the world wars, Hercules expanded, with a focus on developing both gasoline and diesel engines to address the growing challenges faced by trucking and heavy equipment operations.

The Hercules production line was further expanded to meet the quantity of engines required during World War II. The company, with a payroll of more than five thousand people, manufactured over 750,000 gasoline and diesel engines for Allied military vehicles, ships and various equipment. Hercules provided engines and power plants for every conceivable type of mechanized vehicle, including tanks, armored cars, scout cars, tank transporters, landing craft, jeeps, amphibious tractors, landing craft and trucks for every purpose. Following the conclusion of hostilities, Hercules could not adjust to the postwar market changes. In other words, World War II was the company's gold era for production output, revenue and profits.

A foundry workman at Hercules Motors Corporation manually pouring molten steel into an engine block mold, circa 1942. *Library of Congress.*

In 1956, Hercules unveiled a new line of three-, four- and six-cylinder diesel and overhead-valve gasoline engines that were interchangeable. These engines were engineered with identical rods, valves, cylinder blocks and crankshafts for cost savings, particularly regarding engine maintenance and parts inventory.

Cleveland-based Hupp Corporation acquired Hercules in 1961, making the Canton plant into its Hercules Division. Five years later, White Motor Company purchased the Hercules Division of Hupp Corp. The Canton plant was then labeled the White Engine Division. This division was sold as a separate entity in 1976, and the company was completely reorganized under the new name White Engines Inc.

The new company mainly supplied U.S. Defense Department contracts in building engines for military vehicles, such as multi-fueled engines for two-and-a-half-ton and five-ton trucks, and engines for fifteen- and thirty-kilowatt generator sets. White Engines had a private brand contract with Caterpillar Tractor to supply engines for its "Tow Motor" line. White also engineered a seventy-horsepower diesel engine especially for the Ford Motor Company for use in its Ford E-350 line of trucks.

A stationary generator is powered by a custom-built Hercules diesel engine, circa 1944. *Library of Congress.*

After being sold in several deals that failed to be profitable, White Engines was acquired by a group of investors, many of whom were former Hercules officers. The company was then renamed Hercules Engines Inc. and, at the time, maintained a full-time staff that exceeded six hundred employees.

Hercules stayed in business by securing a series of contracts for military trucks in the United States and with NATO countries abroad. The company also continued to supply engines to equipment manufacturers such as fork companies, but even this was only marginal. The 1990s proved to be a time of unstable cash flow and barely breaking even with cost. Eventually, the military contracts stopped being awarded altogether. The six-hundred-thousand-square-foot, Canton-based motor plant ceased all production operations and closed its doors in 1999, marking the end of another great contributor to Ohio's manufacturing heritage.

10

HEAVY INDUSTRY GETS HEAVIER

Jeffrey Mining Company and Galion Iron Works

In 1876, Francis Marion Lechner (1838–1915), with the help of a banker named Joseph Andrew Jeffrey (1836–1928) and a group of other investors, incorporated the Lechner Mining Machine Company to manufacture coal-mining machinery. This collaboration occurred after Jeffrey witnessed the demonstration of a preliminary model of a mining machine invented by Lechner. The purpose of Lechner's invention was to significantly decrease the time required during the coal-mining process known as the under-cutting. Traditionally, this required a miner to lie on their side for uncomfortable, endless hours, digging out a hole or what is called a "kerf" that measures six inches thick and four feet deep and could vary from ten to twenty feet in width. This was the under-cutting operation, and the only tools the nineteenth-century miner had were a pick, a hoe-like tool to pull out the picked-out coal and a hand shovel. The under-cutting operation was regarded as the most time-consuming and dangerous task in mining coal. Lechner's answer to alleviate this backbreaking labor was an engineered, pneumatic, chain driven, air-powered machine that could under-cut coal mechanically.

Lechner projected that his machine could cut between six hundred and eight hundred tons of coal daily. In 1877, the under-cutter machine prototype was completed. It weighed 1,350 pounds and was powered by a compressed-

Jeffrey Manufacturing Company coal-cutting equipment being used in a West Virginia coal mine, circa 1925. *West Virginia University Library.*

air motor (electric motor technology had not yet been developed). On completion, the Columbus shop couldn't supply enough air pressure to test the device. But Francis Lechner proceeded to ship the equipment to a New Straitsville, Ohio coal mine in which Joseph Jeffrey had a financial interest. Once the under-cutter was unloaded, it was moved into the mine and near the air compressor. The air pressure reached 50 pounds and was increased to 80 pounds, then finally to 115 pounds. The activated machine made one revolution and then flipped off the truck, landing upside down. The under-cutting machine was taken back to Columbus, where Lechner had a variety of so-called technical experts diagnose and attempt to troubleshoot the problem, to no avail.

Eventually, with perseverance, the problems of the cutter were worked out and the business started taking orders from coal mines, but Jeffrey had serious concerns regarding the business's financial health. In 1882, Lechner stepped down as general manager, probably because Jeffrey, who was now company president, refused to allot large amounts of money to further inventive research. By 1887, the company was operating well beyond the breakeven point and was finally realizing excellent returns. Jeffrey resigned his part-time position with the bank, bought out the other shareholders, including Lechner, and increased his own salary to $5,000 per year (equal to $160,000 in 2023). He also changed the firm's name to the Jeffrey Manufacturing Company and relocated the business to a four-acre site on First Avenue. By 1888, the company was firmly in the business of manufacturing electric motor-driven coal cutters, electric underground locomotives, elevators and material conveyors. The underground electric locomotive for transporting miners and coal soon became Jeffrey's second-largest product line.

In the 1890s, the company began marketing internationally. The cutting machines and locomotives were being sold and shipped to mines in England, Wales, Australia and South Africa. In 1904, the Jeffrey Manufacturing Co. acquired the Ohio Malleable Iron Company to assure a source of malleable iron castings needed to make the steel chain associated with its entire equipment line. By 1908, a third of the sales revenue was from the chain- and materials-handling divisions.

By 1914, the company was occupying forty-eight acres on First Avenue and employed over 4,500 workers. Jeffrey was now the world's largest manufacturer of coal cutters and mining and industrial locomotives. The company site had a private telephone system with nearly two hundred

The Jeffrey Manufacturing Company mine rail transport system, circa 1920. *West Virginia University Library.*

phones, along with a transportation system consisting of electric trucks. The site was even serviced by its own water utility. In regard to employee welfare, there was a cooperative store that stocked groceries, provisions and clothing. Jeffrey had one of the first industrial infirmaries in the country, staffed by a physician and nurses. To help employees in the event of sickness or accident, a mutual aid association was established. There was also a loan association to help employees finance buying a home.

The record year for sales was 1920, and in 1926, the company took steps to acquire the Diamond Coal Cutter Company Limited of Wakefield, England. The most important acquisition that proved lucrative for the Jeffrey family was the purchase of the Galion Iron Works and Manufacturing Company in Galion, Ohio. Jeffrey paid $2 million in 1929 for Galion, which was considered twice what the company was worth. But by 1940, and in the wake of the Roosevelt administration's programs such as the Work Projects Administration, that initial investment and more had come back in Galion dividends. And that was just the start. During the Great Depression, the Jeffrey Columbus plant was operating one or two days a week, but the Galion works were operating overtime, because highways were in great demand. Eventually, the company's sales of heavy construction equipment exceeded the sales of mining machinery by more than threefold. In time, Galion became the world's largest manufacturer of road rollers and second to Caterpillar in motor graders.

Galion Iron Works was established by David Charles Boyd (1864–1932) in 1907. Initially, Galion manufactured a wide range of heavy road-building and other construction equipment, such as drag scrapers, wagons, plows, stone unloaders and rock crushers. By 1911, Galion started fabricating a light-duty, horse-drawn road grader.

Due to the wide range of equipment produced and the company's success, it was reorganized in 1923, and the name was changed to the Galion Iron Works and Manufacturing Company. The "Light Premier" was an early grader manufactured in 1915. It was advertised as light enough for only two horses but strong enough for four. Its blade could be raised, lowered, tilted, angled and shifted sideways, as seen on modern graders.

Galion had a good reputation for producing some of the largest pull-type graders in the heavy-equipment industry. These huge machines were pulled by the largest traction engines and crawler tractors available and easily outperformed other motor graders of the day. Galion discontinued selling its pull-type graders in 1945. In 1922, Galion developed an innovative self-propelled motor grader. This equipment was arranged with the transmission

The Galion Iron Works manufactured rollers and earth graders used to build America's twentieth-century roads and highways, especially during the WPA era. Photograph circa 1922. *Library of Congress.*

and tractor engine in the rear of the frame, while the operator cockpit was located near the center of the machine. In addition, Galion's engineering staff developed one of the company's greatest achievements, the Galion hydraulic control. The successful application of hydraulic technology was implemented on both pull-type and self-propelled graders.

Jeffrey's profitable status returned again in 1934, and in 1936, the company introduced the first universal cutting machine, known as the 29U. It was the coal-mining industry's backbone until after World War II. The majority of Jeffery's product line was regarded as vital to the war effort. The company enjoyed favorable quotas and priorities for materials during this time.

Internationally, the Jeffrey Manufacturing Company and its subsidiaries employed approximately 7,500 workers. The British Jeffrey-Diamond subsidiary manufactured special radar equipment for detecting German aircraft. Jeffrey Manufacturing won five U.S. Navy "E" excellence awards for supplying the chain for the ammunition hoists on a majority of the navy's destroyers and cruisers built after 1942. A new building was constructed especially for this purpose; the building is now a part of the State Library of Ohio.

In 1946, sales of the company's subsidiaries represented 38 percent of the total revenue, and the percentage figure also corresponds to the amount of profit. The company structure was reorganized, and the original company name was changed to the Jeffrey Company.

While the company's postwar performance continued to slide and losses were incurred, British Jeffrey Diamond and Jeffrey-Galion were firmly in the black. From 1951 through 1962, these two company subsidiaries' total net income represented 80 percent of Jeffrey's net income. But in 1974, Jeffrey Mining Products was sold to Dallas-based Dresser Industries Inc.

MARION SHOVEL

The Marion Steam Shovel Company was founded in 1884 in Marion, Ohio. The company manufactured steam shovels. Business boomed during the late nineteenth century as railroad construction occurred in the American and Canadian West. With the United States' acquisition of the Panama Canal Zone in 1903, the federal government turned to the Marion Steam Shovel Company to provide the steam shovels necessary for the canal's construction.

Marion Steam Shovel in service building the Panama Canal, circa 1908. *Library of Congress.*

Top: The Marion Steam Shovel Company industrial campus, circa 1920. *Marion Public Library*.

Bottom: The Marion Power Shovel Company built the NASA crawler to transport rockets like the *Saturn V* to the launch pad, as shown here circa 1985. *National Museum of the U.S. Air Force*.

Marion, Ohio, became known as the "city that built the Panama Canal," thanks to Marion Steam Shovel.

The Marion Steam Shovel Company continued to prosper during the 1920s and 1930s. In the 1920s, the company manufactured the largest shovel to that time. As steam power became less popular, the Marion Steam Shovel Company changed its name to the Marion Power Shovel Company. In the 1930s, the firm manufactured an even larger steam shovel. This one weighed approximately 1,500 tons; forty-six railroad freight cars were required to transport the shovel to its destination. In the twentieth century, the Marion

Power Shovel Company's shovels were used primarily for roadbuilding and strip-mining. Its shovels also were used in constructing both the Hoover Dam and the Holland Tunnel.

The Marion Power Shovel Company also assisted the United States in the exploration of outer space. The company manufactured the crawling transporters used to carry the Apollo rockets to the launch pad. The various Space Shuttles continue to use the haulers today to travel to the launch pad.

In 2003, a Wisconsin company purchased the Marion Power Shovel Company. At its peak, Marion Power Shovel employed 2,500 workers, but as of 2003, the firm had just 300 employees.

BATTELLE MEMORIAL INSTITUTE

The will of Gordon Battelle established the Battelle Memorial Institute in 1923. Battelle is seen here about 1919. *Columbus Metropolitan Library.*

Battelle Memorial Institute Inc. is the world's oldest and largest independent charitable trust organization devoted to scientific research and technical development. The institute is a worldwide operation with corporate headquarters in Columbus, Ohio. Battelle's mission involves working to identify, assess, modify and apply new and existing technologies for both private and government clients in a wide range of disciplines and fields. These areas of study include but are not limited to the health sciences, space travel, manufacturing, computer technology and environmental research. More specifically, Battelle has been instrumental in the development of such diverse products and processes as no-melt chocolate, plastic six-pack straps, liquid correction fluid, the universal product code (UPC) system and cruise control for automobiles. In one year alone, the institute was known to have been involved with over 5,500 different projects for nearly 1,700 different clients.

The Battelle Memorial Institute was financed and created as a result of an entire family passing away within a period of seven years. This began with the death of Ohio steel industrialist Colonel John Gordon Battelle (1845–1918), who left his fortune to his wife and only son. In 1920, his son Gordon Battelle (1883–1923) had a will prepared that, on his death, would authorize the creation of the Battelle Memorial Institute. In 1923, at the age of forty, Gordon died unexpectedly following a routine appendectomy. He left nearly half of his estate to establish the proposed institute. Two years later, his mother, Annie Maude Norton Battelle (1865–1925), also died of a heart condition. Her will bequeathed the balance of the family steel and mining fortune to the institute. In all, the institute began with an endowment of $3.5 million ($57 million in 2023).

Gordon Battelle's will specifically stated that the purpose of the institute was to concentrate on "education in connection with and the encouragement of creative and research work and the making of discoveries and inventions in connection with the metallurgy of coal, iron, steel, zinc and their allied industries." The Battelle Memorial Institute was incorporated as a nonprofit corporation in 1925, and construction of its headquarters was started on a ten-acre parcel of land adjacent to Ohio State University.

In 1928, while construction of the headquarters building was underway, the chairman of the board of trustees realized that the Battelle family endowment was in steel stocks that were inflated in value. The board voted to liquidate all of the stocks and invest in government bonds. If this had not taken place, the original endowment would have been lost in the October 1929 stock market crash. Ironically, the institute's first laboratory successfully opened in October 1929 with a staff of about thirty employees. The institute's first director was the metallurgist Dr. Horace W. Gillett (1883–1950). Research work in the 1930s was centered on analyzing the properties of steel, cast iron and copper; the innovative uses of pulverized coal; and the durability of nickel.

During World War II, the federal government turned to Battelle for research into the composition and propulsion of missiles and rockets. In 1943, Battelle scientists began studying a metal called uranium on behalf of work being done on the Manhattan Project in Oak Ridge, Tennessee. Today, the institute manages several of the country's nuclear research centers. Its research on peacetime nuclear reactors and propulsion led to the development of the nuclear submarine *Nautilus* in 1955.

One of Battelle's better known technological developments is xerography. Inventor Chester Carson (1906–1968) was issued a patent for electrophotography on October 6, 1942. He approached the institute in 1944. Working through a Battelle subsidiary, the institute initiated an agreement to assist Carson in refining his electrophotography technology. Both Battelle and Carson devoted the next five years and the funding to developing this new technology, which was based on static electricity.

Chester Carlson, seen here about 1965, invented a dry-printing process called electrophotography. With the assistance of Battelle, he helped transform his invention into Xerox. *Columbus Metropolitan Library*.

Battelle also began a lucrative relationship with the Haloid Company, a struggling photographic paper manufacturer. Haloid entered into an agreement with Battelle to assume financial cost of future xerography research in exchange for the opportunity to commercialize the technology. Battelle then traded its xerography patents to the cash-poor Haloid Company for equity in the company. In 1961, the company changed its name to Xerox Corporation.

In addition to managing and operating its own research facilities, Battelle also manages, and in some cases, co-manages, a number of national laboratories on behalf of the U.S. Department of Energy. Here are some of the facilities:

1. Brookhaven National Laboratory (co-managed in a collaboration between Battelle and Stony Brook University).
2. Los Alamos National Laboratory (co-managed in a collaboration between Battelle, the University of California and the Texas A&M University System).
3. Oak Ridge National Laboratory (co-managed in a collaboration between Battelle and the University of Tennessee).
4. Pacific Northwest National Laboratory.
5. Savannah River National Laboratory (through the Battelle Savannah River Alliance).
6. National Biodefense Analysis and Countermeasures Center on behalf of the Department of Homeland Security.

Ohio-based Battelle Memorial Institute is first and foremost a nonprofit global organization that applies science and technology to work at creating and sustaining a much-needed safer, healthier and hopefully more secure world.

STEEL IN OHIO

Early nineteenth-century iron production in Ohio usually took place using a furnace located in isolated rural communities established on property owned by an iron company. All of the materials necessary to produce iron—coal limestone, timber and iron ore—were available in the immediate region. After these raw materials were exhausted, the furnace operation was shut down and relocated to a new area with an ample supply of resources. Most early furnace operations produced pig iron, which was used to cast and fabricate components to make machinery, building supplies, tools and a wide range of kitchen items. Heavy iron forges were also established to produce a higher quality of iron.

One of Ohio's first iron-making operations was the Hopewell Furnace, established in 1804 near Youngstown. But during the first half of the nineteenth century, southern Ohio dominated the iron industry. By 1860, southern Ohio had sixty-nine iron furnaces producing more than one hundred thousand tons of iron annually across Vinton, Jackson, Hocking, Scioto and Gallia Counties. The furnace owners would send most of the pig

With access to the Great Lakes, a larger volume of coal and iron ore could be delivered from the region to steel mills in Cleveland, Youngstown, Toledo and Akron. Photograph circa 1920. *Cuyahoga County Public Library*.

iron to Cincinnati and Pittsburgh via the Ohio River to be fashioned into finished products. The iron producers in the southern part of the state relied on charcoal to fuel their furnaces. This eventually left the regional landscape mostly bare of trees, which were needed to make charcoal. The supply of iron ore also declined following the American Civil War. This resulted in the decline of the economic prosperity of southern Ohio.

Northeastern Ohio emerged as the major region for iron production in the second half of the nineteenth century. One reason was access to iron ore reserves from the Great Lakes region. During the 1840s, charcoal-fired furnaces began being replaced with furnaces fired by coal in northeastern Ohio. Coal furnaces produced iron that was cleaner and of higher quality. The population in northeast Ohio quickly grew because of the increase in iron production and the discovery of abundant coal resources. Business owners such as John D. Rockefeller and Samuel Mather located their production refineries and mills to Cleveland, Youngstown, Canton and Akron because of the ample supply of coal and iron in the region. An example of this industrial and commercial movement is that Cleveland's population jumped from 17,000 people in 1860 to 160,000 by 1880. Iron production and coal mining allowed Ohio to emerge as one of the most prosperous and progressive industrial states in America by 1900.

ARMCO

The American Rolling Mill Company (ARMCO) was organized in 1899 in Trenton, New Jersey, by industrialist George Verity (1870–1942) and a group of investors. The next year, construction of the steel mill was underway in Middletown, Ohio. The mill was placed in service and produced its first tonnage of steel in 1901. In Cincinnati, Verity was already in charge of the American Steel Roofing Company. He originally planned to consolidate the two steel production operations into one comprehensive steel mill in order to produce the final product. But the Middletown Board of Trade convinced Verity to relocate the business from Cincinnati to

A fully operational wide-strip continuous rolling mill produces steel coils, circa 1955. *Cincinnati & Hamilton County Library.*

Middletown. From the beginning, ARMCO produced rolled sheets of steel stock, despite the fact that the early production methods proved to be difficult and very labor-intensive.

In 1904, ARMCO allowed the formation of a shop committee to give workers a means for entering into dialogue with management regarding working conditions. This concept of labor relations was one of the first in Ohio. It was similar to a union but didn't require workers to pay dues.

In 1921, John Butler Tytus, (1875–1944) invented the first practical wide-strip continuous rolling process for manufacturing steel. This process greatly reduced the cost of making steel and was first implemented in a new ARMCO plant in 1924. It transformed the American steelmaking industry. The traditional process produced about 510 tons of steel material monthly, while the Tytus process easily rolled out 40,000 tons a month. As a result, ARMCO became one of the most successful steel producers in the United States.

In 1978, the company's name was changed to Armco Steel Corporation in order to reflect its non-steel acquisitions. The company's headquarters remained in Middletown until 1985, when it moved to New Jersey, and subsequently to Pittsburgh, Pennsylvania, only to return to Middletown in 1995. In 1999, Armco was merged with AK Steel Corporation. At the time of the merger, Armco employed approximately 5,700 employees and had plants in Mansfield, Coshocton and Zanesville, Ohio; and Butler, Pennsylvania. Today, AK Steel Corporation is headquartered in Butler County, in Ohio's West Chester Township and is a subsidiary of Cleveland-Cliffs.

CLEVELAND-CLIFFS

Over the last two centuries, Ohio steel manufacturers have come and gone. But there is one company that ranks best in class as a steel producer. This is Cleveland-Cliffs, North America's largest flat-rolled steel producer.

Cleveland-Cliffs is a vertically integrated, high-value steel firm. The company has the unique advantage of being self-sufficient. It is able to carry production from the extraction of raw materials such as iron ore through the manufacturing of steel products, tubular components and stamping and tooling. Additionally, it produces carbon and stainless-steel tubing products, hot and cold stamped components, die design and tooling. The company also offers iron ore pellets, hot-briquetted iron and coking coal. The company operates steel mills and finishing facilities in Ohio, West Virginia, Kentucky, Indiana, Illinois, Michigan, Pennsylvania and North Carolina with an annual rate of production of about twenty-three million net tons of raw steel. The company also owns and operates six iron ore mines in the Great Lakes region that yield various grades of iron ore pellets, including standard and fluxed, used in blast furnaces as part of the steelmaking process.

The company began as the Cleveland Iron Mining Company and was established in 1847 by attorney Samuel Mather (1817–1890) and six other Ohio-based associates. They started the business after learning about the high-quality iron ore deposits recently discovered in Michigan's Upper Peninsula. Soon after, the first Soo Locks opened, in 1855, allowing iron ore to be shipped from Lake Superior to Lake Erie.

A number of technological advancements, such as the British Bessemer process, enabled the North American Great Lakes region to produce steel on an expanded industrial scale. Lake Erie's south shore had access to an abundant supply of coal shipped by rail from southern Ohio, West Virginia and Kentucky. These factors made the region the perfect place to establish steel mills.

Looking to the future, in 2020, Cleveland-Cliffs acquired AK Steel Corporation and ArcelorMittal USA holdings. Following the acquisition

The Bessemer process, illustrated here about 1890, enabled the Great Lakes region to produce steel on an expanded industrial scale. *Cuyahoga County Public Library*.

of these two prominent steel companies, along with the completion of the Toledo Direct Reduction plant, Cleveland-Cliffs is strategically positioned to maintain its product quality and industry prominence through the next century. Cleveland-Cliffs recognized the need for industrial manufacturers everywhere to be better stewards of our environment. The company is committed to reducing waste in general, improving water conservation and reducing carbon emissions by 25 percent by 2030 and has issued a promise to become North America's leader in steelmaking and mining sustainability.

YOUNGSTOWN SHEET AND TUBE COMPANY

The Youngstown Sheet and Tube Company of Youngstown, Ohio, was a steel manufacturer established in 1900 by Colonel George Dennick Wick (1854–1912), who served as founding president, and James Anson Campbell (1854–1933), who served as secretary and later chairman. In time, the Youngstown Sheet and Tube Company became one of America's most important steel producers. It should be noted that the company's president, George Wick, was returning from Europe when he perished in the Atlantic on April 15, 1912, during the sinking of RMS *Titanic*. James Campbell was head of the company during the period of labor strife that included the January 1916 East Youngstown labor riot that required the National Guard to intervene to restore order. In 1922, East Youngstown was renamed Campbell by the community in order to distance itself from the memory of the infamous riot. During World War I, Campbell was director of the American Iron and Steel Institute, a position that made him responsible for the allocation of steel for the war effort. In 1923, the company purchased the Brier Hill Steel Company, also of Youngstown. That year, the plant facilities of the Steel and Tube Company of America were also acquired. Together, these two acquisitions made the company the fifth-largest steel producer in the country and the Mahoning Valley's largest employer.

The works in Campbell and Struthers, Ohio, were the company's main plants and were equipped with four blast furnaces, two Bessemer converters, twelve open-hearth furnaces, several blooming mills and a slabbing mill. Other heavy equipment included seamless tube mills, a butt-weld tube mill, a 79-inch hot strip mill and 9-inch and 12-inch bar mills at the Struthers Works. The Brier Hill Works was also very well equipped, with two blast furnaces named Grace and Jeannette, twelve open-hearth furnaces, a 40-

Workmen are carefully pouring molten steel into molds at the Youngstown Sheet and Tube Company's Campbell Works, circa 1935. *Mahoning County Public Library.*

inch blooming mill, a 35-inch intermediate blooming mill, a 24-inch round mill, an 84-inch and a 132-inch plate mill and an electric-weld tube mill. During the Depression years, the Brier Hill Works sat idle until it reopened in 1937, and most of the plant's production was for the Campbell seamless tube mills.

After World War II, the postwar boom saw industrial workers asserting a higher expectation to increase their standard of living. Collective bargaining didn't always result in a mutual benefit or agreement. This often resulted in a walk-out labor strike. The late 1940s and early 1950s was a time that saw an expanding American economy in the mist of the Korean War, which placed steel products and production in high demand. In 1952, President Harry S. Truman tried to seize steel mills in the United States to prevent a strike. This led to the U.S. Supreme Court ruling in the case of *Youngstown Sheet & Tube Company v. Sawyer*, which limited the president's authority in such matters.

The use of foreign over domestic steel introduced a new type of competition that was difficult to confront, especially for older, large-scale, unionized steel manufacturers. The company abruptly closed its Campbell Works and dismissed five thousand full-time workers on Monday, September 19, 1977. The Brier Hill Works and the company's steel mills in Indiana were sold to Jones and Laughlin Steel. The doors of Brier Hill were permanently closed in 1979 as part of a growing trend of steel mill closings that completely devastated the Youngstown community and economy. The Youngstown Sheet and Tube Company, which employed over twenty-seven thousand people in 1950, was no more.

HOBART BROTHERS

In 1914, Charles Clarence Hobart (1855–1932) left Hobart Manufacturing Company, a firm he started seventeen years prior, and established the Hobart Brothers Company. His wife, Lou Ella, and their three sons, Edward, Charles and William, were also active in the company. The firm first manufactured a variety of products, including generators, metal office furniture and air compressors. It began fabricating electrical apparatus that included motors and battery chargers.

The company decided to manufacture its first line of electrical welders in 1925. This decision placed Hobart Brothers on the road to becoming a preeminent company in the welding industry. Also, with the opening of the Hobart Institute of Welding Technology in 1930, the company became a prominent supporter of welding education. The school began operating as a separate, nonprofit entity in 1940 and has over the years trained more than one hundred thousand trade welders.

In 1937, Hobart Brothers started producing welding stick electrodes. With World War II on the horizon, this proved to be excellent timing in preparation for an active role in wartime production. Hobart Brothers manufactured over one hundred thousand welders and approximately forty-five thousand generators in support of the war effort. Immediately after the war, the company received the Army-Navy "E" Award for excellence for its wartime contribution.

American Airlines asked Hobart Brothers in the mid-1940s to design a generator specifically for starting large aircraft engines. The result was what came to be known as the Hobart Ground Power generator. By 1958, Hobart Brothers initiated manufacturing solid welding wires and then turned to the development and production of tubular welding wires in the mid-1960s.

The Hobart Brothers Company was instrumental in training and certifying female welders during World War II. Workers are seen here about 1943. *Library of Congress.*

Both were marketed under the Hobart brand. In 1986, Hobart Brothers acquired two companies, adding metal-cored welding wire development to its capabilities. Other acquisitions have allowed Hobart Brothers to remain an industry leader for future decades. In 1995, the family-owned and operated Hobart Brothers was acquired by Illinois Tool Works, but it is still headquartered in Troy, Ohio.

11
THE MAKING OF TOYS, SHOES AND MATCHES

RAINBOW CRAFTS COMPANY

In Cincinnati, Ohio, Joseph McVicker (1929–1992) was working for his mother's company, Kutol Products, which made soap and wallpaper cleaner. In 1955, McVicker discovered that the firm's wallpaper cleaner was also suitable to be used a malleable modeling clay. After successfully testing the claylike substance at several local elementary schools and daycares, McVicker began selling it at a Washington, D.C. department store under the name Play-Doh.

Joseph McVicker discovered that Kutol brand wallpaper cleaner was also an excellent modeling clay. He packaged and sold it as Play-Doh. Photograph circa 1960. *Courtesy Hasbro Inc.*

In 1956, Joseph McVicker and his uncle Noah McVicker (1905–1980) established the Rainbow Crafts Company strictly to manufacture and market Play-Doh. The new company offered the product in multiple colors and in eleven-ounce containers. In 1965, after only nine years in business, Rainbow Crafts Company Inc. and all of the rights to Play-Doh were acquired by General Mills.

In 1971, General Mills decided to merge the Rainbow Crafts division into Kenner. Play-Doh was now officially part of the Kenner line of toy products. In 1985, General Mills consolidated both its Kenner and Parker Brothers toy divisions to establish a new subsidiary, Kenner Parker Toys Inc. Two years later, Kenner Parker was acquired by Tonka, which maintained Kenner as a separate division. But in 1991, Hasbro Inc. acquired Tonka and, in 2000, closed the Cincinnati Kenner offices after merging that company's product line into its own.

KENNER

Kenner, a toy company based in Cincinnati, Ohio, was established in 1946 by brothers Albert, Phillip and Joseph L. Steiner. In 1958, they began to use television as an advertising tool to market their toys throughout America. During this time, Kenner created a corporate mascot, known as the Kenner Gooney Bird, which featured in its television ads and in their company logo,

Kenner introduced the iconic Easy-Bake Oven, invented by Ronald Howes. The toy is seen here circa 1963. *Courtesy Hasbro Inc.*

"If It's Kenner! It's Fun!" The Bubble-Matic, a toy bubble-blowing gun, was one of Kenner's original toy products.

Other popular Kenner toys were Girder and Panel building sets, introduced in 1957 and designed with girls as well as boys in mind; the Give-a-Show projector, created in 1959; and the iconic Easy-Bake Oven, which was invented in 1963 by Ronald Howes. It sold for $15.95 ($150.00 in 2023). After only eleven years in business, Kenner was purchased by General Mills in 1967.

Ohio Art Company

In 1908, the Ohio Art Company was established in Archbold, Ohio, by a dentist named Henry S. Winzeler (1876–1939). Winzeler saw a bigger future in manufacturing a variety of novelty items. After selling his dental practice, he used those funds to rent a space and hire fifteen workers to fabricate picture frames that were then shipped throughout the country, to Canada and to Mexico. Business was good, and soon the operation required a larger shop space. The Bryan, Ohio Chamber of Commerce, coupled with additional funding from local citizens, provided the resources to entice the Ohio Art Company to relocate to their community in 1912 and build a new factory. The company was now in a position to grow and was capable of manufacturing twenty thousand metal picture frames daily. The company purchased metal lithography equipment that enabled the firm to produce advertising signs, scale dials and small toy wagons. This was the start of decades as a profitable toy maker.

The importation of toys from Germany to America was completely curtailed with the outbreak of World War I, and this proved a grand opportunity for American toy manufacturers. Business rapidly increased as Winzeler added to his line of metal lithographed toys and toy components. During this time, a popular line of tea sets was introduced, followed by colorful sand pails and shovels in 1923. In 1932, the Walt Disney Company granted Ohio Art one of the first licenses to produce one of its toy characters, Steamboat Willie. In the 1950s, plastic toys started replacing metal lithographic toys, and Ohio Art adapted to meet the trend by investing in Champion Molded Plastics.

In 1978, the Diversified Products Division was formed to coordinate the production and marketing of Ohio Art's metal lithography capabilities.

The front facade of the headquarters of the Ohio Art Company, circa 1990. *Courtesy Ohio Art Company*.

Ohio Art's biggest twentieth-century triumph was the Etch-A-Sketch, invented by a French electrician and first sold in 1960. *Courtesy Ohio Art Company*.

Examples of products made were Kodak's film canisters, popular novelty signs and also TV trays. Other popular brands that were licensed include Coca-Cola, Budweiser and Campbell's Soup. Ohio Art enjoyed an internationally recognized reputation for superb quality, and this was one of the driving forces behind its continued success.

One of Ohio Art's bigger triumphs was a toy trade-named Etch A Sketch, which premiered in July 1960. Etch A Sketch was a unique drawing toy invented by French electrician Andre Cassagnes (1926–2013). The product was introduced in 1959 at the International Toy Fair in Nuremburg, Germany. The Ohio Art Company initially thought the advance payment was too high and negotiated a reasonable amount months later with an associate of Cassagnes's. Originally, Cassagnes referred to the toy as L'Ecran Magique, which is still the toy's name in France. The Ohio Art Company launched the toy in America in time for the 1960 Christmas season under the name Etch A Sketch. Following its introduction that year, Ohio Art sold over six hundred thousand units of the drawing toy at a retail price of $2.99 ($29.00 in 2023). Over one hundred million units were ultimately sold worldwide.

The Ohio Art Company no longer manufactures toys, and it ceased manufacturing the Etch A Sketch because, in 2016, all rights to the name and design were sold to Canada-based Spin Master. However, the company is still in the business of precision printing and high-quality metal lithography. Today, Etch A Sketch is a registered trademark of Spin Master.

Ohio Shoe Manufacturing

Portsmouth, Ohio, was regarded as the "Shoe Capital of the World," as it played a major role in the American shoe manufacturing industry. Robert Bell (1815–1883) traveled to Portsmouth in July 1850, and the following month he opened the first shoe manufacturing factory in town. Bell is regarded as the founding pioneer of the shoe and boot industry in Portsmouth.

Shoes had always been hand-crafted until Bell decided to invest in a shoemaking machine in 1869. That same year, the factory employed fifteen women and twenty-five men and had a daily output of two hundred pairs of shoes. Bell left most of the daily management of his factory to his foreman, Frederick Drew. This gave Bell time to focus on starting a wholesale shoe business with partners W.H. Ware and Joseph Vincent.

Jan Ernst Matzeliger (1852–1889) was the inventor of the shoe lasting machine. He is shown here circa 1885. *Library of Congress*.

But Bell had his sights on other pursuits and exited shoemaking entirely by 1874 to pursue a career selling insurance. In 1873, Bell's son-in-law, along with two others, formed the Portsmouth Shoe Manufacturing Company. It factory stood on Front Street. The company's listed stockholders were Fredrick Drew, Irving Drew, George Paden and Henry Paden.

Four years later, in 1877, Irving Drew, the son of Robert Bell's foreman, Fredrick Bell, established the Irving Drew & Company as a shoe factory in Portsmouth. Two years later, Irving Drew partnered with George Selby to reorganize the firm as Drew, Selby and Company. In 1881, the factory relocated to a larger leased building with a boiler. Shoes previously made using only manual labor were now made in a new factory powered by steam.

The international shoe industry was revolutionized at this time with the introduction of a device developed by a thirty-one-year-old Black inventor named Jan Ernst Matzeliger (1852–1889). Matzeliger invented the shoe-lasting machine, and this radically changed how shoes were manufactured. The process of making shoes by hand required a skilled shoemaking to replicate the form and size of a foot by creating a wooden mold called a last. This was then used as the pattern to size and shape the shoes. Manual assembly of the soles to the upper shoe required well-crafted skill to tack and sew the two parts together. Prior to Matzeliger's invention, such intricate work was feasible only if executed by trained hands. This gave the shoe laster substantial control within the shoe industry, similar to that of an exclusive club or union.

It took Matzeliger five years to perfect his automated shoe laster. He was granted a patent in 1883. Normally, a skilled hand laster could complete fifty pairs of shoes working a ten-hour day. Matzeliger's invention was easily capable of finishing between two hundred and six hundred pairs of shoes a day. This volume of manufacturing reduced by half the cost of shoes for the consumer and considerably increased the profit margin for the manufacturer.

By 1891, Drew, Selby and Company purchased an entire city block and began building a state-of-the-art factory incorporating Matzeliger's automated shoe laster machines. In 1906, Drew sold his interest in the firm to Selby and went on to open a shoe factory under his own name, Drew

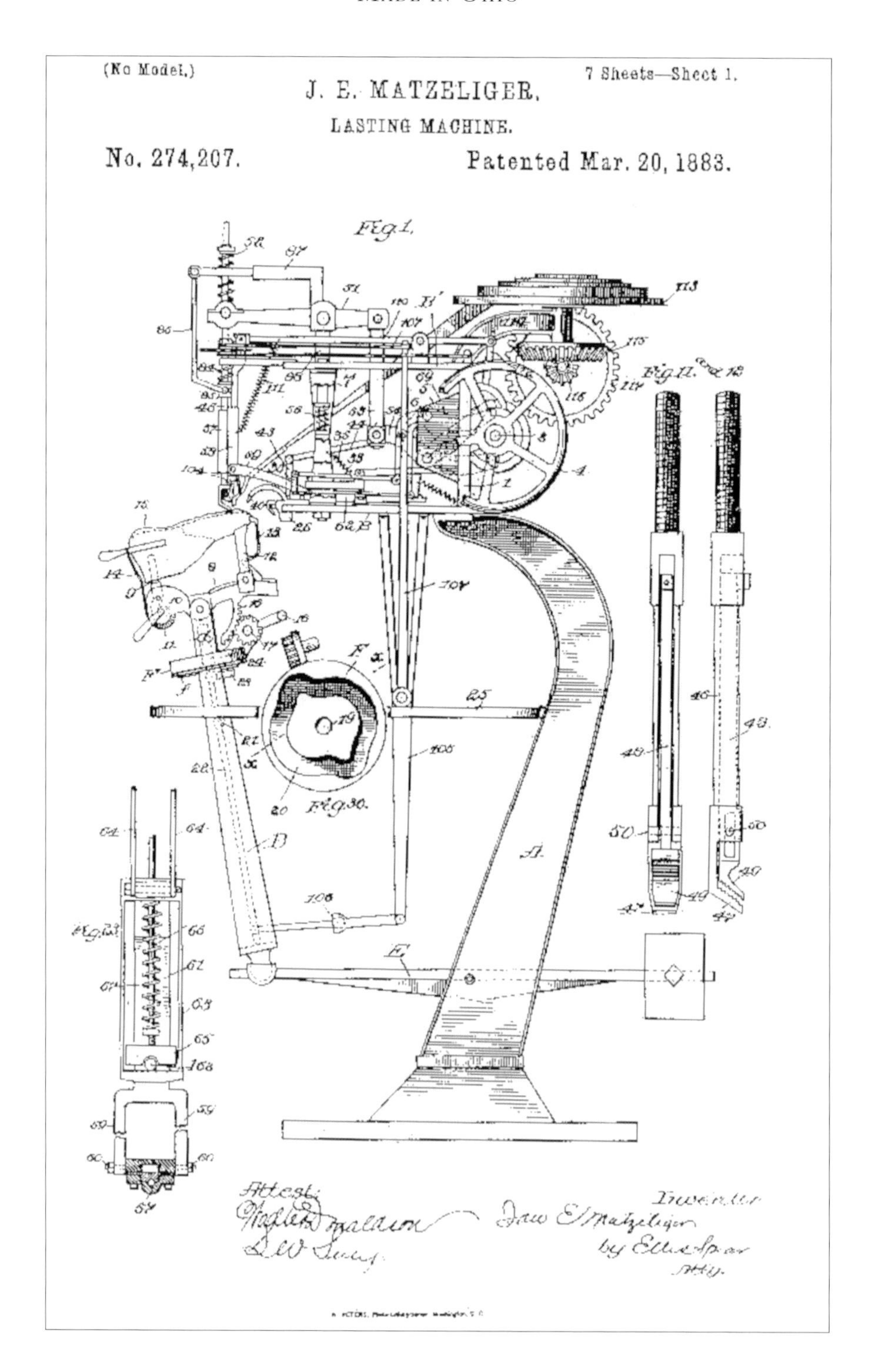
(No Model.)
7 Sheets—Sheet 1.
J. E. MATZELIGER.
LASTING MACHINE.
No. 274,207.
Patented Mar. 20, 1883.
Fig. 1.
Fig. 30.
Attest:
Inventor
by Ellis Spear
Atty.

Opposite: An 1883 patent document for the Matzeliger shoe lasting machine, which increased shoe manufacturing productivity. *Library of Congress.*

Above: Two employees of Selby Shoe Company in Portsmouth, Ohio, working at making shoes, circa 1935. *Shawnee State University Clark Memorial Library.*

Shoe Corp. After the 1937 Ohio River flood, the Drew Shoe Corporation uprooted itself from Portsmouth and relocated to Lancaster, Ohio. There, Drew Shoe occupied a building previously operated by the H.C. Godman Shoe Company of Columbus. Selby also revised the corporate name to Selby Shoe. By 1927, Selby was employing a staff of more than two thousand workers, and his firm was the eighth-largest shoe manufacturer in the world. Because of the demand, Selby shipped its shoes to Australia, Canada, England and throughout the world.

Other shoe manufacturing companies that operated in Portsmouth included Excelsior Shoe Manufacturing Co.; Paden Brothers & Co.; Russell, Vincent & Williams; Star Shoe Co.; and Tremper Shoe Co. Established in 1920, one of the leading makers of women's shoes in America was Williams Manufacturing Company. Its shoes were reasonably priced, which was a major factor during the years of the Great Depression.

Columbus, Ohio, was a major manufacturer of shoes in the nation. Shoe manufacturing was a major industry in the city between 1880 and 1975. In 1900, it was estimated that one of every eight shoes made in the United States was produced in Columbus. The Columbus shoemaking industry began in 1849 and was a little above a cottage industry at best. But 1880 saw the cottage trade of shoemaking upgraded when H.C. Godman (1832–1907) began manufacturing lavish, high-class shoes for consumers of the post–Civil War industrial age. Many Americans at that time had the general impression that quality shoes could not be manufactured west of the Alleghenies. But Godman and his high-quality line of shoes changed that thinking.

Brothers Robert F. and Harry Wolfe were employed by Godman. By 1890, they had struck out on their own and, in 1901, established the Wolfe Brothers Shoe Company. Operating a fully mechanized factory, they were able to manufacture seven thousand pairs of shoes a day. Profits from the company were later invested to acquire two Columbus newspaper publishing houses, the *Ohio State Journal* in 1903 and the *Columbus Dispatch* in 1905. The *Columbus Dispatch* is the metro community's primary newspaper and was owned by the Wolfe family until its sale in 2015. In addition to Godman and Wolfe, other major shoe manufacturers in Columbus included the Riley Shoe Co., Kropp Shoe Co., Fenton Shoe Co., Bradford Shoe Co. and Walk-Over Shoe Company.

DIAMOND AND OHIO MATCH COMPANIES

The Diamond Match Company has its roots in several nineteenth-century East Coast companies. In 1853, Edward Tatnall (1782–1856) of Wilmington, Delaware, was working to develop a match that would not ignite with the slightest friction. Tatnall's answer to this problem was to decrease the phosphorus content by 25 percent. He set up a company to produce his matches but died in 1856, and his shop never reopened due to the financial depression of 1857.

In 1861, two of Tatnall's former business associates, Henry Courtney and William H. Swift, established the Swift & Courtney Company. They manufactured a line of new, safer matches under the trade name Diamond State Parlor Matches. Over the next two decades, the company was acquired or merged with a number of other match manufacturers until 1880, when

In the nineteenth century, women were generally employed to do the delicate work of making matches and were paid based on daily productivity. Workers are seen here circa 1870. *Akron-Summit County Library.*

all assets were acquired by Ohio Columbus Barber (yes, that was his actual name). Barber was the owner of the Barber Match Company of Akron, Ohio. Taking full advantage of the already established product trade name, Barber renamed the firm the Diamond Match Company. He relocated the company factory to Barberton, Ohio, immediately after the panic of 1893. At that time, Barberton was an industrial town that Barber was developing to house his factory employees and a variety of local suppliers. By 1900, the Diamond Match Company was the largest match manufacturer in America.

The Ohio Match Company of Wadsworth, Ohio, was at one time the second-oldest match manufacturer still in production in the nation. Originally established by E.J. Young in 1895 to produce wooden matches, the factory was eventually tooled to also turn out book matches. The Wadsworth plant produced more than three hundred million wooden and paper matches a day. The Wadsworth facility was a maze of brick structures and covered eighteen acres. The company at one point operated additional plants in Colorado, New York, California, Washington State and Missouri. Due to the simple and straightforward nature of the business Ohio Match wasn't one to embrace innovation. This was most likely because of the world demand for its product. However, its product line did include famous trademarks, including Ohio Blue Tip, Ohio Noiseless, Ohio Safety Matches and Royal Star.

In 1928, the Ohio Match Company was acquired by Diamond, which then turned around and sold a one-half interest to the Swedish financier, entrepreneur and industrialist Ivar Kreuger. But Kreuger's international match empire collapsed in 1932 following his death. Diamond managed to reorganize Ohio Match in 1936, and by 1948, the company had emerged once again as an independent business entity. The company was now one of the three largest match producers in the world. Ohio Match became a subsidiary of Hunt Foods in 1957; a decade later, Hunt Foods was merged with Norton Simon Inc. After enduring several additional acquisitions both domestic and foreign, the Ohio Match Company fell victim to the post–industrial era collapse that began to take a serious toll on American manufacturing in general. The Ohio Match Company factory in Wadsworth closed its doors in 1987. But the Diamond brand name still survives and is America's leading match manufacturer. Today, the Diamond brand of matches is owned and manufactured by Royal Oak Enterprises LLC., which produces over twelve billion matches annually.

12

GROCERY GOODS, JELLY AND SOAP

KROGER

The Kroger chain of grocery stores and supermarkets was established in Cincinnati, Ohio, by pioneer grocer Bernard H. Kroger (1860–1938) and has a 140-year history of feeding a sizable portion of America from coast to coast. After ten years of learning from his father, who once owned a grocery store that went bankrupt, and also working as an unappreciated grocery clerk and manager, Bernard Kroger decided it was time to strike out on his own in 1883. He established a grocery store in downtown Cincinnati using his own funds of $372 (the equivalent of $12,000 in 2023). On July 1, 1883, Kroger opened the Great Western Tea Company, and by the end of his first year in business he was operating four stores. The Kroger Grocery and Baking Company was incorporated in 1902 with a chain of forty grocery stores with Bernard Kroger serving as president.

Obsessed with having the lowest prices, Bernard Kroger was among the first large-scale retailers to eliminate the middle man, in 1901 opening a bakery to supply bread for his stores and three years later purchasing a butchering and packing facility to supply meats. His company was also among the first to hire female cashiers and to own its own fleet of trucks for supplying stores, instead of contracting with local truckers.

Traditional nineteenth-century food shopping involved patronizing the classic butcher, baker, vegetable/fruit vendor and dry-goods mercantile.

Top: The Kroger Grocery and Baking Company was incorporated in 1902 with a chain of forty stores. Photograph circa 1900. *Cincinnati & Hamilton County Library*.

Bottom: Kroger early on delivered meat, groceries and baked goods to local residents. One of its delivery wagons is shown here circa 1895. *Cincinnati & Hamilton County Library*.

Thinking of his customers' convenience, Kroger sought to consolidate as much as possible into an innovative, one-stop shop. In 1901, he established an in-house bakery and meat-cutting/packing facility to supply his stores. This was an innovative first in the nation that reduced costs by eliminating middle distributors.

Eyeing the merits of making his own products, Kroger once cost-effectively purchased a large quantity of fresh cabbage from a local farmer. He then enlisted his mother's help in making delicious German sauerkraut using her own recipe. The Cincinnati German community wasted no time buying the crocks of sauerkraut.

Mary Kroger's kitchen set the stage for large-scale, private-label food manufacturing in the nation. Private-label goods represent approximately a quarter of the company's total sales. Kroger manufacturing plants include dairies, bakeries and facilities for producing general grocery and dry-goods items. Kroger owns and operates dairy plants such as Centennial Farms in Atlanta, Georgia; Crossroad Farms in Indianapolis, Indiana; Heritage Farms in Murfreesboro, Tennessee; Mountain View Foods in Denver, Colorado; Swan Island Dairy in Portland, Oregon; Tamarack Farms in Newark, Ohio; and Vander Voort's Dairy in Fort Worth, Texas. It operates eleven additional dairies in eleven other states.

Its general grocery and dry-goods production facilities include State Avenue of Cincinnati for salad dressings, broths, red sauces, jams, jellies and syrups; Springdale Ice Cream & Beverage in Springdale, Ohio, for soft drinks, ice cream and bottled waters; Kenlake Foods in Murray, Kentucky, for nuts, cornmeal and hot cereal; and Pontiac Foods of Pontiac, South Carolina, for coffee, sauces, seasonings, noodles and rice. The company has two additional general grocery plants in two other states.

Kroger also owns and operates industrial bakery plants, including Country Oven Bakery in Bowling Green, Kentucky; Anderson Bakery of Anderson, South Carolina; and RCK Foods in Kenosha, Wisconsin. It operates six other dairies in five other states. Kroger pioneered self-service shopping to the public in 1916, eliminating the need to wait for groceries to be delivered.

Today, the Kroger Company is America's largest retail grocery chain, with over 2,860 stores in thirty-five states doing business under twenty-eight different names, including QFC, Dillons, Fry's, Ralphs, Fred Meyer, Smith's, Home Chef and Vitacost (e-commerce). On average, 238 million prescriptions are filled annually through Kroger's 2,255 in-store pharmacies. Over 1,545 supermarket locations have Kroger gasoline fueling centers, which offer a discount over many major oil companies. This is all a part of the founder's original intent to allow customers to realize more savings during their shopping experience.

The Kroger company owns and operates a sizable fleet of trucks, trailers and refrigeration trailers to supply stores in lieu of contracting with

Bernard H. Kroger, seen here about 1920, established himself as a pioneer grocer in Cincinnati, Ohio, beginning in 1883. *Cincinnati & Hamilton County Library.*

commercial trucking companies. Kroger is also testing and investigating the merits of driverless vehicles for grocery delivery. Ever the pioneer, in 1972, Kroger started testing the use of electronic scanning for retail grocery inventory control. Today, the company uses the QueVision checkout program, which has proven effective in facilitating customer checkout and reducing long lines.

Today, Kroger is addressing the issue of sustainability by instituting a program to collect safe, edible, fresh food products and promptly redistribute them to local food banks. Kroger has also established partnerships with local organizations such as Cincinnati's Last Mile Food Rescue to redistribute surplus and unused food of institutions and restaurants to areas where food is in short supply. Kroger is truly an innovative Ohio company where ingenuity has been used to better serve the welfare of the public on a daily basis.

QUAKER OATS COMPANY

The pioneering miller Ferdinand Schumacher (1822–1908) was born in Hanover, Germany. In his youth, he apprenticed in the grocery business and made oatmeal by using medieval milling technology. In 1851, at the age of twenty-eight, he immigrated to America with his brother Otto. They planted roots in Akron, Ohio, in 1852 and opened a grocery store stocked with simple, inexpensive items that the public could afford. These items included whole oats, which Schumacher decided to grind as he had in Germany and sell to the public as breakfast food. Unfortunately, the locals regarded oats only as feed for livestock. Addressing this dilemma, Schumacher started searching for alternate ways to prepare whole oats. In 1854, he started shaping the oats into small square cubes that weighed only one ounce. Schumacher's new way of processing whole oats presented the public with an easy way to prepare and use oats as a table food. Consequently, the demand for oat field crop production began to increase.

From this humble start, the idea of eating oatmeal spread throughout the rest of the nation. In 1856, Schumacher purchased a mill and later leased water power on the Ohio Erie Canal in Akron to power his mill that produced oatmeal. In 1858, equipment for pearling barley grain was installed, along with special equipment that increased an employee's productivity to a daily output of twenty barrels of ground oats. The plant continued to be expanded, and in 1875, steam power was placed in service.

During the Civil War years, demand for oatmeal increased exponentially, as the federal government purchased oatmeal to feed Union troops in the field. Ohio's excellent transportation systems of canals and railroads allowed Schumacher to expand his regional markets. During the war, demand for oats was so great that, in 1863, Schumacher had to relocate the milling operation to Mill Street in Akron, and there he started the Empire Barley Mill. At the Mill Street location, Schumacher continued researching other methods for processing oats. This work led to the development of a process that involved precooking the oats. After drying, they would turn into small flakes. Schumacher's flaked oats were quickly accepted by the public and became a bestseller.

But in 1886, disaster struck when the Empire Barley Mill was completely destroyed by fire. As a man of faith, Schumacher believed that the Divine would keep him and his property safe from harm. He thus saw no reason to purchase insurance. This was a major setback, not just for Schumacher but also for many Akron residents, as Schumacher's mill was the major local employer. Also at this time, Schumacher's was coming up against competition from other oat processors, such as the Akron Milling Company. Even with competition, Schumacher's product line was regarded by the public as superior to that of other brands. Since he did not have the financial capital to rebuild his mill, Schumacher presented a proposal to the Akron Milling Company. His plan called for merging his company with Akron Milling. Akron Milling's president understood that Schumacher's reputation for producing a quality product meant a substantial return in profits and agreed to the merger. This was the beginning of the F. Schumacher Milling Company.

The Quaker Oats Company was established by pioneering miller Ferdinand Schumacher, seen here circa 1875. *Akron-Summit County Library.*

A Quaker Oats Company advertisement from 1910. *Akron-Summit County Library.*

Also referred to as the "Oatmeal King," Schumacher introduced oatmeal to American households as a viable meat substitute that was inexpensive, filling and nutritious. Schumacher's largest mill, the so-called Jumbo Mill in Akron, sold 360,000 pounds of oatmeal a day, and his recipe for breakfast oatmeal was included in the first cookbook to contain recipes for oatmeal, in 1873. Although Schumacher was the most prominent oatmeal producer in the market, several smaller yet prosperous mills cropped up throughout Ohio and the Midwest. In Ravenna, Ohio, Henry Parsons Crowell managed a water-powered milling company called Quaker Oats, where he oversaw all stages in manufacturing oatmeal, from grading and cleaning to packaging and shipping. Stamped with the iconic Quaker mascot, Crowell's Quaker Oats was the first registered trademark for a cereal brand.

Robert Stuart, another well-known oatmeal producer, had a prosperous oatmeal operation in Chicago with his father. While Crowell and Stuart were successful oatmeal manufacturers, their success couldn't touch the production and popularity of Schumacher's products. Schumacher joined the Oatmeal Millers Association and regained his wealth and influence as president of the American Cereal Company, a consolidation of the seven largest oatmeal mills in the country. In 1901, the Quaker Oats Company was formed by the consolidation of Schumacher's German Mills American Oatmeal Company and four other successful oat milling companies.

Over the decades of its success, Quaker Oats has produced a variety of breakfast and grain-based products, such as Chewy granola bars. In 2001, Quaker Oats was bought out by PepsiCo, as the beverage company was enticed by Quaker Oats' ownership of Gatorade, which Quaker Oats had purchased in 1983.

J.M. SMUCKER COMPANY

Orrville, Ohio's very own J.M. Smucker Company is the nation's premier producer and seller of jams, jellies and preserves. The company operates internationally and manufactures and markets ice cream toppings, peanut butter, syrups, coffee and pet food and pet snacks. The firm is currently under the fifth generation of Smucker family leadership.

The company was founded in 1897 by Jerome Monroe Smucker (1858–1948) in Orrville, Ohio. In 1897, he constructed a cider mill in Orrville. The company's original supply of apples is thought to have come from trees planted locally in Orrville by early nineteenth-century lay clergyman John Chapman, better known as Johnny Appleseed. Smucker prepared apple butter and sold it from the back of his horse-drawn wagon. Smucker made the apple butter from his Pennsylvania Dutch great-great-grandmother's recipe. By using steam-heated copper coils, Smucker's eldest son, Willard, employed a secret method to capture the vapors usually lost in cooking. This gave the spread a very unique flavor that soon drew an excellent following. Smucker staked his reputation on each crock of apple butter with his personal signature on every lid.

Under Willard's leadership, the company incorporated in 1921 and by 1928 was distributing preserves and jellies in Ohio, Pennsylvania and Indiana in such quantities that the Pennsylvania Railroad installed a special side spur to the Smucker plant to expedite the delivery of ingredients.

Jerome M. Smucker, founder of the J.M. Smucker Company, was the patriarch who established the company's standard of quality. He is seen here circa 1920. *Courtesy J.M. Smucker Company.*

In 1935, Willard opened a Smucker's pre-processing apple plant in the state of Washington that shipped apples to the Orrville plant for additional preparation. He also was looking to switch to a modern glass package without losing the Smucker reputation for "old-fashioned" quality. The new crock was developed by an employee at the facility. A local art student was hired to design the label, which depicted a pioneer lady boiling a kettle of apple butter over an open fire. The new glass jars proved very successful and became a trademark in 1939.

In 1940, Smucker's came out with its first line of ice cream toppings and within two years was distributing nationally. In 1946, Federal USDA Agriculture inspectors were compensated to oversee every aspect of its production operation. As a result, Smucker's received a "U.S. Grade A Fancy" designation. This standard of quality justified Smucker's higher markup and better shelf placement in stores.

Smucker's third generation of leadership began when Willard's son Paul joined him in making strategic plans for future diversification and expansion. Like his father, Paul had started working in the family business at a young age. Paul graduated from Miami University in Oxford, Ohio, with a business degree in 1939 and then started full-time with the family business as a cost accountant.

The company received merger offers from food industry giants like Borden and Quaker Oats. But, in 1959, it offered a third of the company's stock to the public to raise funds for capital investments. In 1961, Paul was made president of Smucker's and continued to fiercely protect the company's product and corporate image. From its beginning, the company remained focused on maintaining comprehensive control of production.

In 1960, production was increased by 40 percent with the opening of a manufacturing plant in Salinas, California. The company's product offerings

The original cider mill in Orrville, Ohio. The Smucker's Company's original supply of apples is believed to have come from early nineteenth-century trees owned by John "Johnny Appleseed" Chapman. Photograph circa 1897. *Courtesy J.M. Smucker Company.*

grew to exceed one hundred varieties, including ice cream toppings, which by 1960 accounted for 20 percent of sales.

Willard Smucker was not a big fan of advertising, but in 1961, the company retained Wyse Advertising of Cleveland to develop radio ads. The agency came up with the slogan, "With a Name like Smucker's, It Has to Be Good." Bolstered by the catchy new tagline, from 1962 to 1965, sales substantially increased.

In 1974, Smucker's focused on becoming the country's number-one jelly manufacturer. With increased advertising and more market penetration, particularly in the South, Smucker's overtook Welch's for the top spot among jelly and jam manufacturers, with over one-fourth of the market, by 1980.

The fourth generation of family leadership began with Tim and Richard Smucker. Tim became vice-president of planning in the 1970s, and this move ushered in a more modern approach to strategic, tactical and operational planning. After Tim became president and chief operating officer, a four-year capital improvement and plant expansion program was announced in 1982. This followed a 39 percent year-to-year gain in earnings. Richard served as treasurer, vice-president, executive vice-president, chief administrative officer, chief financial officer and company president. It should be noted that the company had no CEO following the death of Paul Smucker in 1999. But beginning in 2001, Tim and Richard served jointly as co-CEOs until 2011, when Richard became the sole CEO.

Smucker's began international acquisitions in the late 1980s. The firm entered the 1990s with 38 percent of the domestic jam and jelly market. The fourth-generation leadership set a higher goal of capturing a 70 percent market share. This meant taking on experienced rivals like Kraft, Welch's and strong local store brands. To accomplish this, in 1994, Smucker's ended its thirty-five-year relationship with Wyse Advertising in favor of Leo Burnett Co., an agency that company executives felt could help meet the goals they were setting for their brands.

In the early 1990s, members of the Smucker family were still the majority stockholders. In the food industry, consolidations were gaining momentum, so the family implemented a new class of nonvoting stock to reinforce against any aggressive takeover strategy. From 1990 to 1993, sales continued to increase.

The J.M. Smucker Company became part of the Standard & Poor's 500 Index in 2008. That year, the company purchased the Knott's Berry Farm food division from ConAgra Foods, the Canadian milk brand Carnation and the Folgers coffee brand division from Procter & Gamble. In addition to

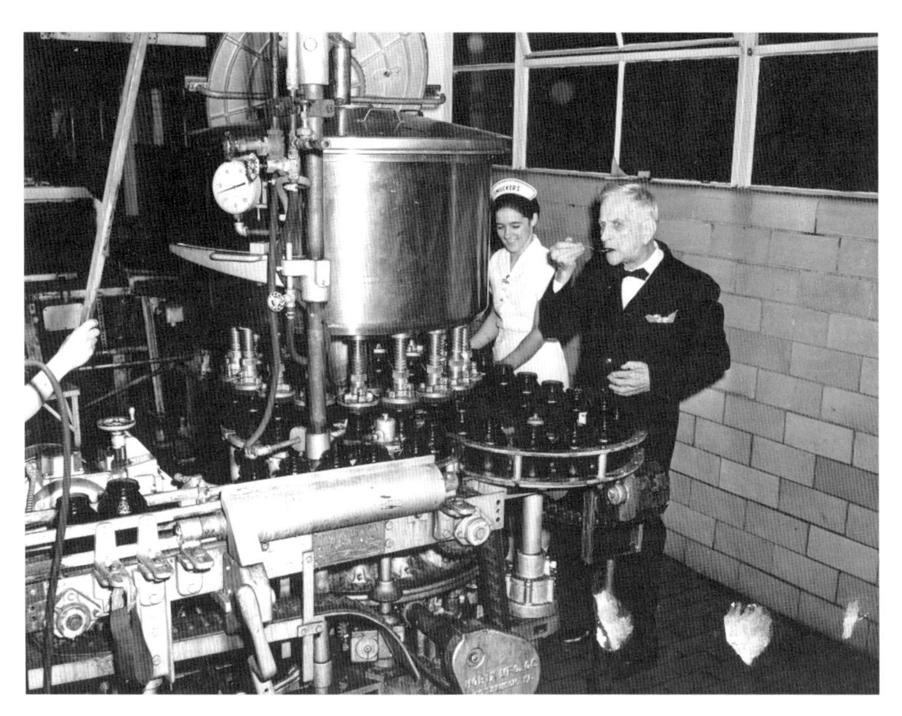

Mr. J. M. Smucker sampling a taste of product from the production line, circa 1940. *Courtesy J.M. Smucker Company.*

adding several other coffee brands to the company's product line, Smucker's acquired an impressive list of pet food brands beginning in 2015.

With the appointment of Mark T. Smucker as CEO in 2016, Smucker's top leadership role transferred to the family's fifth generation. Under his leadership, the company's communications and marketing has been focused on interacting much more closely with consumers in general.

Mark Smucker has emphasized the company's goal of increasing corporate awareness of social responsibility, environmental protection and sustainability. With Mark Smucker at the helm, the company has already achieved its major goals, ahead of expectations. One goal was to divert and recycle 95 percent of waste, which was previously earmarked for a landfill. Another goal was to reduce water usage by 15 percent. The company's greenhouse gas emissions have been decreased by 10 percent, reducing its carbon footprint. J.M. Smucker Co. is truly a fifth-generation example of an innovative "Made in Ohio" industry leader.

Procter & Gamble

Ohio's very own Procter & Gamble Company (P&G) is now an international manufacturer of consumer goods and household products with production facilities in thirty-six countries. Its categories of products consist of home care, baby care, family care, fabric care, beauty care, health care, beverages and snacks.

William Procter (1801–1884) was a candlemaker, and James Gamble (1803–1891) was an accomplished soap maker. The men had settled in Cincinnati and married sisters. Taking the advice of their father-in-law, these two men pooled their talents and money in 1837 to establish their own company, Procter & Gamble. The company prospered with the growth of America until, by 1859, sales had reached the $1 million mark ($33 million in 2023).

The business employed a staff of eighty full-time workers just prior to the Civil War. During the war, P&G was a major supplier of soap and candles to the Union forces. Profits increased significantly during the war years. Military service personnel in general were introduced to Procter & Gamble products, and the P&G logo of the moon and stars became a symbol of quality with Union soldiers. Procter & Gamble expected wartime shortages. So, to maintain full production, it had to find and develop new methods of

William Procter, candlemaker, partnered with James Gamble, soap maker. They were the founders of P&G in 1837. The men are seen here circa 1860. *Cincinnati & Hamilton County Library*.

manufacturing. Until the second year of the Civil War, stearic acid, used to make candles, was a by-product of lard stearin, which became expensive and was in short supply. At this time, it was discovered that tallow could also be used to produce the stearic acid.

In 1875, the first full-time chemist was hired to work with James Norris Gamble (1836–1932), who was also a chemist, to develop new products. Their collaboration produced an inexpensive floating white soap that was equal in quality to expensive castile white soap, which also floated. In 1878, Procter & Gamble's White Soap entered the market and placed the company at the forefront of the industry.

An interesting characteristic of the product was developed accidentally after a worker left a soap mixer on while taking a lunch break. The increased mixing time resulted in more air being added to the batch of ingredients. Before long, the company was receiving large orders for its "floating soap," and the name of the product was called Ivory Soap.

Procter & Gamble continued to expand and evolve and was renowned for its progressive work practices in the late nineteenth century. William Procter's grandson William Cooper Procter (1862–1934) instituted profit-sharing in 1887. He reasoned that if the staff had a stake in the company, they would think twice before striking.

As electricity and electric lights became more common in households, P&G turned its attention to focusing on manufacturing over thirty different types of soaps by the 1890s and began decreasing candle production. By 1920, P&G decided to curtail manufacturing candles. Company research began turning out new products. In 1911, the company introduced Crisco, a cooking shortening made of vegetable oil instead of high-saturated animal fats. Camay body soap was introduced in 1926, and Oxydol joined the P&G product line of laundry detergents in 1929.

As radio became commonplace in the 1920s, the company began sponsoring radio programs that particularly appealed to homemakers. These programs soon became known as soap operas. By 1939, P&G was sponsoring twenty-one radio broadcasts. The advertising section did the same with great success when television became a common fixture in homes in the 1950s.

Throughout the twentieth century, Procter & Gamble moved to establish itself as a multinational corporation and branched out into new areas. With the start of the postwar baby boom, the company introduced Tide laundry detergent in 1946, followed by Prell shampoo in 1950. In 1955, P&G introduced the first fluoride toothpaste under the trade name Crest. With the 1957 acquisition of Charmin Paper Mills, the company began manufacturing toilet paper and various paper products. Also, that year, the Clorox Chemical Company was acquired. An innovative wood pulp drying process brought about White Cloud toilet paper in 1958 and Puffs tissues in 1960. This process also improved the Charmin brand of toilet paper by making it softer.

Looking to the laundry needs of the consumer, P&G introduced Downy fabric softener sheets in 1960. The next year, a disposable diaper product under the trade name Pampers was test-marketed. Prior to this, there were no disposable diapers. The simplicity of Pampers began to replace cloth diapers, which required time and labor to clean. Research continued to improve many products during the 1970s, and new products were added, including Bounce fabric softener in 1972 and Sure antiperspirant and Coast soap in 1974.

P&G entered the over-the-counter drug market with the 1982 acquisition of Norwich-Eaton Pharmaceuticals, makers of Pepto Bismol and Chloraseptic. Its biggest purchase took place in 1985, when it acquired the Richardson-Vicks Company, maker of Vicks respiratory care products, NyQuil cold remedies, Oil of Olay skin care products and the motion-sickness treatment Dramamine. From G.D. Searle & Company, P&G acquired Metamucil laxative. These acquisitions made Procter & Gamble a leader in over-the-counter drug sales.

A special soap developed and manufactured by Procter & Gamble that floated was trade-named Ivory and is available in stores today. Image circa 1875. *Cincinnati & Hamilton County Library.*

During fiscal year 1985, P&G experienced its first profit decline since 1953. The mass-marketing practices that had served it so well in previous years lost its effectiveness as broadcast television viewership declined from 92 percent to 67 percent in the mid-1980s. Much of this was a result of cable television.

In September 1988, P&G moved into the cosmetics business for the first time when it acquired Noxell Corporation, the maker of Noxema products and Cover Girl cosmetics. P&G continued to be on the lookout for profitable acquisitions and, in November 2001, bought the Clairol hair-care business from the Bristol-Myers Squibb Company. P&G sold several marginal food business holdings, including selling the Jif peanut butter and Crisco shortening brands to the J.M. Smucker Company in 2002 and Folgers coffee a few years later.

Even though Procter & Gamble is a multinational, Ohio-based corporation dedicated to making a profit for its stockholders, it should be stated that for over 180 years its goal has also been to improve the quality of life of the world's general population. P&G's Good Everyday program, launched in May 2020, has as its goal to bring people together and do good for the world in areas such as environmental responsibility and sustainability, equality and inclusion. Procter & Gamble continues to expand globally but is still a company on the river in Cincinnati, Ohio.

13

NEW MACHINES FOR THE HOUSE

Hoover Company

Canton, Ohio department store janitor James Murray Spangler (1848–1915) invented a portable electric vacuum cleaner that he first tested in 1907. After several modifications, it was patented in 1908. Spangler then established the Electric Suction Sweeper Company to manufacture his invention.

William Henry Hoover (1849–1932) and his son Herbert W. Hoover Sr. (1877–1954) were in the business of manufacturing leather goods in North Canton, Ohio. William Hoover's wife was Spangler's first cousin and had purchased one of Spangler's early models. After she showed the machine to her husband and son, William Hoover decided to invest in Spangler's company and purchased the sweeper patent. Spangler died in 1915, and Hoover became president of Electric Suction Sweeper Company. In 1922, the company name was changed to the Hoover Company and H.W. Hoover became president.

Under Hoover's direction, several improvements were made in the design and engineering of the vacuum cleaner. New marketing strategies were tested. Early advertisements offered the consumer a free ten-day, in-home trial to test the vacuum cleaner. Hoover also negotiated an agreement with stores to demonstrate the cleaner to interested customers. Those stores received a commission on any sales, and this established the dealerships. Hoover's innovative marketing strategies and improved technology made the

One of James Spangler's original Electric Suction Sweeper Company vacuum cleaners, sold by William Hoover. Photograph circa 1915. *Akron-Summit County Library.*

A group of women participating in a Hoover sale demonstration class on October 6, 1919. *Akron-Summit County Library.*

Hoover Company the largest vacuum cleaner manufacturer in the world. In addition, the company continued to engineer innovative attachments that made the vacuum cleaner even more appealing to consumers.

Hoover business interests were not just focused on American markets. A Canadian factory in Windsor, Ontario, was opened as early as 1911, and another plant was established in Hamilton, Canada, in 1919, which shipped to England, Ireland and Wales. Sales were so good that eventually a headquarters was established in Perivale, England, in 1932.

William Hoover passed away in 1932, but his company continued to be innovative and to prosper. With the start of World War II, the company retooled its manufacturing facilities to help the war effort. The company manufactured plastic helmet liners and parachutes for fragmentation bombs. It was during the war years that the Hoover family decided to offer stock in the company, going public in 1943 in preparation for a postwar economy. The Hoover Company as a family business was now a publicly traded firm.

The Hoover Company was acquired in 1985 by the Chicago Pacific Corporation, which was then acquired by the Maytag Corporation in 1989. Maytag was eventually acquired by Whirlpool, which in 2007 sold the Hoover floor-care line to Techtronic Industries (TTI). Hoover products are still manufactured with the internationally recognized red circle logo.

WHITE SEWING MACHINE AND WHITE MOTOR

The White Sewing Machine Company began in 1858 and was for a brief time located in Templeton, Massachusetts. It was established as the White Manufacturing Company by founder Thomas H. White (1836–1914), who invented and patented a small hand-operated, single-thread sewing machine that he manufactured and sold as the New England Sewing Machine. After the Civil War, White relocated his factory to Cleveland, Ohio, in 1866. George W. Baker and D'Arcy Porter were two highly skilled mechanics employed by Thomas White. Together, they engineered and patented an improved machine in 1876 that was successfully sold as the White Sewing Machine. The White Manufacturing Company was incorporated that year, and the name was changed to the White Sewing Machine Company. In 1877, a standard model White sewing machine sold to the public for about $60 (equivalent to $1,600 in 2023).

Additionally, Thomas H. White was an automotive pioneer who manufactured cars, trucks, tractors and buses through the later-established White Motor Company. In 1899, he bought one of the first Locomobile steam cars. He was dissatisfied, because the vehicle's boiler proved unreliable. His son Rollin White (1872–1962) decided to improve the steam car's design by engineering a special water tube steam generator, sometimes referred to as a flash boiler. It allowed for better control of steam temperature to produce superheated steam. This allowed the steam car to take advantage of the steam's properties at higher temperatures. In 1900, Rollin White was granted a patent for his steam generator and began selling it to Locomobile. He also persuaded his father to allow him to use space in one of his factory buildings to begin manufacturing automobiles based on his steam generator patent.

Rollin was joined by his brothers Windsor White (1866–1958) and Walter White (1876–1929) in producing an initial group of fifty White Steamers, known as the Stanhope model. These vehicles were offered for sale to the public in April 1901 after having been thoroughly tested. The quality of the steam cars was extremely important, because they were made by a department of the sewing machine company, and White could not risk tarnishing the reputation of his father's company. In 1906, the automotive department was fashioned into its own company, the White Motor Company, with Rollin as vice-president and his father retaining the head position as company president. Production of White Steamers continued until 1911. In 1910, the first gasoline White trucks began production.

President Theodore Roosevelt added a 1907 White Steamer to the White House stable. He permitted the Secret Service to use the car to follow behind him as he rode in a horse-drawn landau carriage. But in 1909, President William Howard Taft had the entire White House stable converted into a garage. Four automobiles were purchased for presidential use, with one of the automobiles being a 1911 White Steamer that cost $4,000 ($122,000 in 2023). The White Steamer was a favorite of President Taft, who was known to use loud bursts of steam against bothersome press reporters and photographers.

The last White steam car rolled off the line in January 1911 as the company began retooling to transition to gasoline-powered vehicles. At the end, approximately ten thousand White steam-powered automobiles were manufactured. This quantity surpassed that of the better recognized Stanley Steamer.

White was successful with its heavy machines and equipment, which saw service around the world during World War I. While White Motor Company

President William Howard Taft and his family riding in the presidential 1911 White Steamer automobile, circa 1912. *Library of Congress.*

remained in the truck manufacturing business for decades, it curtailed all automobile production after the war, focusing entirely on trucks. In short order, the company was selling 10 percent of all trucks made in America. But the Great Depression caused a sever drop in truck sales nationwide. This forced White Motor Company to merge with Studebaker of South Bend, Indiana. Fortunately, White soon regained its independence with the help of government contracts to deliver vehicles for the worldwide war effort.

Prior to World War II, the company was based in Cleveland, Ohio. The White Diesel Engine Division operated out of Springfield, Ohio, and manufactured diesel engine generators, which powered military equipment and camp infrastructure and were used in emergency recovery. After the war started, White designed and was one of several contractors manufacturing the standard U.S. Army reconnaissance vehicle known as the M3 Scout Car. Additionally, White manufactured the later M2, M3, M13 and M16 half-track models. White was producing various sizes of trucks, but following World War II, the decision was made to produce only large trucks, such as semi-tractors.

In the 1950s, White acquired a number of truck manufacturing companies that had previously been profitable building Liberty trucks during the war. These included Sterling and Diamond T. It also bought out Autocar and

White designed and manufactured the standard U.S. Army reconnaissance vehicles known as the M2, M3, M13 and M16 half-tracks. Photograph circa 1942–45. *Library of Congress.*

REO and entered into an agreement with Consolidated Freightways and Freightliner Trucks to sell their trucks through White dealers.

Truck sales were diminishing throughout the 1960s and 1970s. White Motor Company attempted a merger with White Consolidated Industries, but the federal government was blocking the transaction. By the time the government approved a merger, White Consolidated backed out for fear of being hurt by White Motor's financial dilemmas.

By 1980, White Motor Company was insolvent and filed Chapter 11 bankruptcy in the Northern District of Ohio. Swedish automaker Volvo acquired the U.S. assets of the company in 1981, while two energy-related firms based in Calgary, Alberta, acquired the Canadian assets.

FRIGIDAIRE

The Guardian Refrigerator Company was established in Fort Wayne, Indiana, in 1916 by electrical engineer and Dayton, Ohio native Alfred Mellowes (1879–1960) and his investors. Mellowes had developed the first fully self-contained electric refrigerator and wanted to manufacturer and make his innovation available to the public. He set up a small shop and started manually fabricating the electric wood cabinet appliance. Within two years, the company was losing money and nearly bankrupt because of its slow and very low production rate.

General Motors Corp. president William C. Durant could see the merits in manufacturing refrigerators for the post–World War I public. As a result, in 1918, Durant personally purchased the Guardian Refrigerator Company and renamed the firm and its products Frigidaire. The next year, Durant sold his full interest in Frigidaire to General Motors. The entire operation was then relocated to Detroit, where the GM staff instituted mass-production manufacturing processes for the refrigerator. Production issues were resolved, but the unit cost was $775 (the equivalent of $11,500 in 2023). This was still unaffordable to the average consumer.

In 1921, Durant decided to transfer the Frigidaire division to the General Motors Delco Light subsidiary, located in Dayton, Ohio. Under the guidance of electrical engineer and Delco president Charles F. Kettering, Frigidaire quickly made a turnaround in every area of efficiency. With engineering and industrial design upgrades over the next five years, the wooden food cabinet was modified into a steel, porcelain-coated and insulated refrigerator equipped with a temperature control. Frigidaire also introduced a refrigeration line of products to areas where simple iceboxes had previously been the standard method for cooling. Within a decade, Frigidaire refrigeration technology was being applied to ice cream cabinets, milk coolers, refrigerated soda fountains, grocery store display coolers and freezers, drinking fountains and room air conditioners.

By 1926, the manufacturing engineering staff reduced the price of a Frigidaire refrigerator to $468 per unit ($7,500 in 2023). The Frigidaire division was upgraded to a subsidiary of General Motors after surpassing the sales revenues of Delco Light in 1926. Innovative engineering and mass-production efficiencies enabled Frigidaire to produce and sell over one million refrigerators during its first decade with GM.

Despite the Great Depression, Frigidaire refrigerator production exceeded six million by 1941. In the late 1930s, Frigidaire also began diversifying into manufacturing other household appliances, including cooking ranges, ovens, water heaters and clothes washers and dryers. Just prior to World War II, Frigidaire was ranked as the world's largest refrigerator manufacturing plant and had over twenty thousand employees worldwide.

During World War II, refrigerators were considered a public "nonessential" item, and Frigidaire and other appliance makers were drafted into military

Above: The Frigidaire division refined refrigeration technology and offered units affordable to the public. Image circa 1926. *Dayton and Montgomery County Library*.

Opposite: In Dayton during World War II, the Frigidaire subsidiary of General Motors mass-produced Browning machine guns for a third of the original cost. Photograph circa 1942–45. *Library of Congress*.

production. Frigidaire produced a wide range of aircraft parts, including propellers, gas tanks, artillery and bomb hangars. The company also mass-produced more than two hundred thousand Browning machine guns.

The company rolled out its first postwar refrigerator in July 1945 and by June 1949 was back to manufacturing one million units annually. The 1950s saw Frigidaire continue setting innovative standards for the appliance industry with the introduction of automatic icemakers and auto-defrost refrigerators in 1952 and a completely frost-free model in 1958.

The American appliance industry was being besieged by increased foreign imports in the postwar era. In 1960, Edward Reddig (1904–1979) of White Consolidated Industries (WCI), which began as a sewing machine manufacturer, started expanding by acquiring major appliance manufacturers that had lost their competitive edge, such as Philco, Westinghouse and Kelvinator. In 1979, WCI acquired Frigidaire from General Motors.

But in 1986, WCI was acquired by Electrolux AB of Stockholm, Sweden, which owned three hundred companies and is today one of the largest appliance manufacturers in the world. Under Electrolux, Frigidaire was given a new slogan, "Frigidaire Company, Creating a Better Tomorrow." Instead of competing with low-priced foreign imports, Electrolux wanted to differentiate Frigidaire as a desirable brand of high quality. Today, as a subsidiary of Electrolux, the Swedish conglomerate relocated the Frigidaire headquarters from Dayton to Charlotte, North Carolina.

HOBART/KITCHENAID

The KitchenAid brand name was introduced for an electric stand mixer developed by the Hobart Manufacturing Company of Troy, Ohio, in 1919. Image circa 1920. *Dayton and Montgomery County Library.*

The KitchenAid brand name was introduced for an electric stand mixer developed by the Hobart Manufacturing Company of Troy, Ohio, in 1919. In the early 1900s, Hobart started manufacturing the first electric-powered machines for grinding food items, including hamburger, peanuts and coffee beans. In 1915, the Hobart subsidiary Troy Metal Products introduced the first model of an electric mixer designed to mix large quantities quickly. Allegedly, the company executive's wife tested the mixer. Her response was, "I don't care what you call it, but I know it's the best kitchen aid I ever had." As a result, the KitchenAid name was adopted as the mixer's trademark.

During the 1920s and 1930s, the KitchenAid mixer became extremely popular. The mixers were marketed through door-to-door sales and via KitchenAid parties, where a salesperson would demonstrate the appliance features.

New attachments continued to be developed for use in mixing a variety of foods, and Hobart introduced the KitchenAid electric coffee mill, which proved to be very popular. In 1924, Hobart renamed its Troy Metals subsidiary the KitchenAid Manufacturing Company with headquarters in Dayton, Ohio.

In 1926, Hobart acquired an appliance manufacturer that would play an important role in KitchenAid's future. The Crescent Washing Machine Company was established by dishwasher inventor Josephine Garis Cochrane (1839–1913), who originally hailed from Ashtabula County, Ohio, and was the inventor who first patented the automatic dishwasher, in 1886. Cochrane was a manufacturer whom Hobart recognized as a highly successful leader in the

Inventor Josephine Garis Cochrane, shown here about 1885, established the Crescent Washing Machine Company in 1897. *Columbus Metropolitan Library*.

commercial dishwasher market. Research and development for a home unit was delayed because of World War II. But in 1949, Hobart introduced the model KD-10 home dishwasher. It carried the KitchenAid brand name and established a reputation for reliability.

KitchenAid was also known for a range of other products, including instant hot water dispensers, food waste disposers, trash compactors and a line of ovens and stovetops. By 1985, KitchenAid was one of the most recognized and successful manufacturers of home appliances, while Hobart had become one of the foremost producers of commercial and institutional kitchen appliances and equipment. But in February 1986, the Whirlpool Corporation acquired KitchenAid from Hobart. Whirlpool immediately relocated the KitchenAid Ohio staff to St. Joseph, Michigan, to be closer to Whirlpool's Benton Harbor headquarters.

Today, Hobart continues to be a leading manufacturer of equipment products for food preparation, cooking, dishwashing, waste reduction, weighing and packaging. Established in 1897 as the Hobart Electric Manufacturing Company in Troy, Ohio, Hobart has made a dedicated effort to support the needs of the commercial and institutional foodservice and retail food industry by providing state-of-the-art, premium equipment. Its dedication is further demonstrated by having the nation's largest and most experienced service network. Hobart Corporation is a subsidiary of Illinois Tool Works Food Equipment Group.

14

COLORING THE WORLD AND BUILDING WITH BRICKS

SHERWIN-WILLIAMS

The Cleveland-based Sherwin-Williams Company manufactures, distributes and sells a wide range of paints, floorcoverings, coatings and related products to commercial, industrial and retail consumers in over 120 countries.

Starting in 1866, Cleveland bookkeeper Henry A. Sherwin (1842–1916) invested in the paint distributorship of Truman Dunham & Co., but the partnership quickly failed. Sherwin proceeded to establish the Sherwin, Williams, & Co. with Edward Williams (1843–1903) and A.T. Osborn in 1870. The company's first production facility was a former Standard Oil barrel factory, which was purchased in 1873. Normally, a painter would combine paint ingredients to make a finish patch of usable paint. But in 1875, Sherwin-Williams started offering ready-mixed paint that was then sold retail and wholesale through the company's Americas Group, Consumer Brands and Performance Coatings Groups.

The company was incorporated in Ohio in 1884, and in 1886, Osborn sold his interest in the company but retained the retail operations. For the balance of the late nineteenth century and into the twentieth century, the company grew through acquisitions and expansion. In the postwar era of the 1920s, Sherwin-Williams emerged as the largest coatings manufacturer in America. After World War II, Sherwin-Williams introduced a fast-drying interior paint that was water-based and trademarked Kem-Tone.

Henry Alden Sherwin (*left*) and Edward Porter Williams, were the founders of the Sherwin-Williams Company. They are shown here circa 1880. *Case Western Reserve Library.*

Sherwin-Williams Company stores were designed to appeal to the general public as well as the trade painter. A storefront is seen here circa 1895. *Case Western Reserve Library.*

Sherwin-Williams has established a focused Global Supply Chain organization that leverages systems, processes, tools and equipment to deliver best-in-class service for consumers throughout the world. The company operates 137 manufacturing and distribution facilities across five global geographic regions.

The Consumer Brands Group develops, manufactures and distributes a wide range of paints, coatings and related products under a variety of brand names, including Dutch Boy, Cabot, Thompson's WaterSeal, HGTV Home, Krylon, Minwax, Duron, Pratt & Lambert, Purdy and Valspar, to name a few. The Performance Coatings Group supplies a variety of coatings and finishes to automotive and industrial firms, wood furniture manufacturers and packaging and marine markets in more than 110 countries. By 2020, the Americas Group was operating 4,758 retail stores, including over 135 floorcovering centers.

Sherwin-Williams has an international staff of over fifteen thousand employees and plans to maintain its world headquarters in downtown Cleveland, Ohio.

THE BELDEN BRICK COMPANY

The following is a short, detailed history of the Belden family, which has been in the business of manufacturing brick to build this nation for five generations. Like the Smucker family, their fifth generation of leadership is a testament to the strength, diligence and commitment that can work wonders in overcoming and navigating adversity in market climates by applying ingenuity and innovation.

The beginning of the Belden name in brickmaking lies with Henry S. Belden Sr. (1840–1920). While on his father's farm near Canton, in 1873, Henry became interested in the large deposits of coal, clay and shale located there. He began making bricks by the hand-mold process and built a small kiln to study the effects of firing the property's clay and shale. In 1883, he began making paving and fire bricks in a small plant on the farm. The real origins of the Belden Brick Company were in 1885, when Henry assisted in organizing the Diebold Fire Brick Company. In 1895, Henry and three others bought controlling interest in Diebold. The company was appropriately renamed the Canton Pressed Brick Company by Henry Belden Sr.

Henry S. Belden, shown here circa 1900, established the Belden Brick Company and served as president from 1885 to 1920. *Courtesy Belden Brick Company.*

Henry's sons Henry Jr. and Paul joined the company in 1902 and 1904, respectively, beginning the tradition of family involvement. In 1909, L.B. Hartung invested in the company. The association of Beldens and Hartungs continues to the present. Also in 1909, another trend began that would add substantially to the company's growth for many years: the acquisition of other brickmaking companies. The first was when the Canton Pressed Brick Company agreed to take the entire output of the Advance Fire Clay Company of Uhrichsville for a period of five years. In 1911, the trend continued with the acquisition of a small brick plant in Somerset.

A changeover from pressed brick to extruded brick rendered the Canton Pressed Brick Company name obsolete. In 1912, the Belden Brick Company became a reality.

By 1917, Advance Fire Clay Company in Uhrichsville and Enterprise Clay Company in Port Washington had been acquired by the Belden Brick Company. In 1920, the firm lost its founder, Harry Belden Sr. He was succeeded as president by L.B. Hartung, until Hartung's death in 1935. While the original families provided the basis for success, the Belden Brick Company would only be a name without the employees who made it what it is today. Through the years, it has been the superintendents, plant workers and staff who have given form and substance to the company.

One of the first general superintendents was Otto Artzner, who served from 1920 until 1937. Burke Wentz succeeded Artzner as general superintendent. This is the man about whom Bill Belden Sr. said, "When I think about Belden Brick, I think of Burke Wentz—Burke Wentz is Belden Brick." He was instrumental in the expansion of production at older plants, the construction of Belden's first and subsequent tunnel kiln plants and the acquisition and modernization of new plants.

Besides Belden Sr., Burke Wentz had more years of service (fifty-nine) than any other Belden Brick employee. In May 1946, negotiations were completed for the acquisition of the two Finzer Brothers Clay Company

Top: Employees of Belden Brick manually load bricks onto a company rail system, circa 1915. *Courtesy Belden Brick Company.*

Bottom: Classic nineteenth-century beehive kilns are still used to fire bricks in the twenty-first century. *Courtesy Belden Brick Company.*

plants in Sugarcreek. In 1956, the company's first tunnel kiln plant, Plant No. 6, was built in Sugarcreek. A second tunnel kiln facility, Plant No. 8, was completed in 1968. It marked the entry of the company into the soft mud brick market with the Belcrest line.

On February 14, 1970, Paul Belden Sr. died at the age of eighty-seven. The Canton Plant closed that year because of increasing costs and environmental concerns. In 1973, Belden acquired the Shepfer & Moomaw Brothers plant, now known as Plant No. 9, in Sugarcreek. The former Strasburg Brick Company was acquired in 1974. The Somerset, Uhrichsville and Port Washington plants have joined the Canton plant in obsolescence.

In 1979, the company purchased the former Claycraft Brick plant in Sugarcreek. On that site, it built its third tunnel kiln plant. Completed in 1981, Plant No. 3 makes a handformatic, soft mud brick unique to this country. In 1994, the Belden Brick Company Quality Management System was certified as compliant with the ISO 9002 requirements for the Internationally Recognized Standard for Quality Management Systems. This alone set Belden apart from most in the industry. This certification is renewed twice per year as a standard procedure for maintaining this level of excellence and provides the energy required for the objective of continuous improvement.

On September 30, 1996, the Belden Brick Company acquired Redland Brick Inc. (RBI), the U.S. subsidiary of Redland PLC. Redland Brick consists of three plants: Cushwa Brick in Williamsport, Maryland; Harmar Brick in Cheswick, Pennsylvania near Pittsburgh; and KF Brick in South Windsor, Connecticut, near Hartford. In 1998, the Sugarcreek Plant No. 4-1 kilns, the original Finzer Brothers Clay Company plant, were razed. Later that year, William H. Belden Jr., chairman and CEO of the Belden Brick Company, announced that the firm would build a new plant, Plant #2, in Sugarcreek. Operations began in the year 2000 in what is a fully automated brick manufacturing facility with an annual capacity of up to thirty million brick equivalents (BE).

The year 2000 was busy for Redland Brick as it acquired the Rocky Ridge Plant in Maryland from Boral Bricks. Redland also began construction on a new production facility at the Harmar plant to replace an aging facility. Yearly production capacity more than doubled to sixty million BE when the plant became operational in 2001.

In 2007, Belden Brick Plant No. 1, the former Strasburg Brick Company, was shut down. Most of its product lines went to newly revamped kilns at Plant No. 9, making Sugarcreek the sole location of Belden Brick's production facilities. Redland's Rocky Ridge plant was also refurbished with a new kiln and production equipment. Both projects drastically reduced natural gas consumption at these plants.

In 2009, Belden Brick introduced a new product line to the market: thin brick. In response to growing demand for precast panelization of thin brick, Belden Brick constructed a new facility, Plant No. 5, to cut the face

Two employees of Belden Brick work in the molded brick production line of the company. *Courtesy Belden Brick Company.*

off full-size brick units. This unique process allows Belden Brick to be able to offer full-size brick along with perfectly matching thin brick units for a single job. The year 2009 also saw the company vertically integrate its business with the purchase of Sugarcreek's Eureka Machine Shop, founded in 1898.

In 2011, Redland Brick added to its line of brick with the purchase of Lawrenceville Brick in Virginia. These facilities consist of two plants with a capacity of about one hundred million BE. In 2011, the Belden Brick Company was certified as compliant with the ISO 14001 requirements for the Internationally Recognized Standard for Environmental Management Systems. This is another step that sets Belden apart from most in the industry and reinforces their objectives of Environmental Stewardship and Continuous Improvement.

In 2013, Belden Brick Plant No. 9, the former Shepfer & Moomaw Brothers plant, was shut down. Most of its product lines went to Plant No. 8, where there was adequate capacity due to the lingering effects of the Great Recession of 2007–9. In 2016, the molded brick line at Plant No. 8, which made the famous Belcrest brick line, was shut down. The Belcrest line was consolidated into Plant No. 3's line of molded brick

Like many of today's manufacturers, Belden Brick Company has incorporated automation and robotic technology to remain competitive. *Courtesy Belden Brick Company.*

after an investment in new setting equipment. This also prompted a large refurbishment of the fifty-year-old Plant No. 8. This drastically changed its manufacturing method by moving all production to two new extruding lines, drying bricks before they set, installing automated setting capabilities and installing a decorating line to usher in a new era of coated products. The Plant No. 8 project was completed in 2020.

Belden Brick now consists of five manufacturing plants operating eight tunnel kilns, twenty periodic beehive kilns and one brick sawing facility. Belden Brick employs approximately five hundred people and has a total annual production capacity of approximately two hundred million BE. Belden Brick markets more than 250 variations of brick in the United States through a network of 250 distributors and dealers. The firm is generally acknowledged as the industry's quality leader, and has been regarded as "The Standard of Comparison since 1885."

The Belden Brick Company: A Chronology

1885 Henry S. Belden and four associates organize the Diebold Fire Brick Co.
1895 Diebold Fire Brick becomes Canton Pressed Brick Co.
1904 Paul Belden Sr. joins the company
1911 Somerset plant acquired
1912 Canton Pressed Brick becomes Belden Brick Co.
1915 Enterprise Clay Co., Port Washington, acquired
1917 Advance Fire Clay Co., Uhrichsville, acquired
1946 Finzer Brothers Clay Co. acquired
1956 Sugarcreek Plant No. 6 built
1968 Sugarcreek Plant No. 8 built
1970 Canton Plant closed
1973 Shepfer & Moomaw Brothers plant acquired (Plant No. 9)
1974 Strasburg Brick Co. acquired (Plant No. 1)
1981 Sugarcreek Plant No. 3 begins production
1983 William H. Belden Jr., becomes fourth generation to head the Belden Brick Co.
1993 William H. Belden Jr., named chairman/CEO
1994 ISO 9002/1987 certification earned
1995 Robert F. Belden named president
1996 Redland Brick Inc. acquired
1999 Sugarcreek Plant No. 2 built
2000 Redland acquires Rocky Ridge Plant in Maryland from Boral Bricks
2008 Robert F. Belden named president/CEO; Strasburg Plant No. 1 closed; Sugarcreek Plant No. 5 built
2009 Eureka Machine Shop acquired
2011 Redland acquires Lawrenceville Brick, Virginia
2012 Tubar Machine Shop acquired
2013 Sugarcreek Plant No. 9 closed
2016 Robert F. Belden named chairman of the board
2019 Bradley H. Belden named president

15

TWENTY-FIRST-CENTURY PROBLEM-SOLVING OPPORTUNITIES

Looking to the future, Ohio is primed to shake any idea of it being a gone-with-the-wind Rust Belt state to being viewed as a next-generation industrial and technology leader and powerhouse. The twenty-first century is a new era with challenges like global warming, water shortages and economy-stagnating pandemics.

The Columbus Idea Foundry

The Columbus Idea Foundry (CIF) is a unique community of people with creative gifts and inventive natures. As a collaborative community, people are empowered to explore what it takes to make their ideas into a viable and/or marketable product. Since 2008, the CIF has coupled ideas, experience and talents with tools, resources and opportunity to create a shared community of creating and inventing. In addition to being a state of mind, the CIF is a sixty-thousand-square-foot physical facility consisting of workshops, working nooks, offices, classrooms and communal spaces.

The Columbus Idea Foundry's first floor is a space where inventors, artists, artisans, techies and entrepreneurs have access to hand tools, 3D printing and much more. It is an open area where idea creation takes place and prototype models are fabricated. Access to the workshop space is by membership, with an additional hourly rate based on what specific machines are being used.

The Columbus Idea Foundry is a workshop, learning center and creative space for the next generation of inventors. It is a place where people can have access to unique tools and technology. *Courtesy Alex Bandar.*

Prior to using any machine, a member is required to take a class about the machine's operation, safety precautions and material usage. The CIF offers a wide range of equipment orientation classes that cover processes such as laser cutters, welding, CNC plasma cutting and wood turning.

The second floor is a space specifically designed for entrepreneurial and business support to build a company. It has offices, classrooms, conference rooms, an event space, a coworking space and more. The Columbus Idea Foundry community is a diverse group of over eight hundred members, including hundreds of small-business entrepreneurs. It is a place where ordinary Ohio problem solvers can develop their ideas in a climate of ingenuity and networking.

The Coming of Intel

As the United States emerged from the economic stagnation caused by the COVID-19 pandemic, industries found themselves in a vulnerable position due to a critical shortage of microprocessors, also referred to as microchips. The acute global shortage of microchips is the result of an immense demand and a limited supply. Today's computers, automobiles, medical implants, smartphones, electrical equipment and farm machinery operate using some type of microchip. After decades of relying on East Asian microchip production, it has become crucial for the United States to push aggressively to increase its microchip manufacturing capacity.

Intel Corporation CEO Pat Gelsinger has announced that the company will initially invest $20 billion in a new computer microchip manufacturing

Intel has announced that it will be investing $20 billion to build a computer microchip manufacturing facility in central Ohio. *Columbus Metropolitan Library*.

complex in Licking County, Ohio, just east of Columbus. The first phase will involve constructing two microchip plants on a one-thousand-acre site that, when completed, will employ three thousand workers. When the multi-phase Intel Licking County complex expands to its full scale, which is estimated to cost $100 billion, the complex will employ approximately ten thousand workers.

By comparison, the Intel microchip project will be the largest single private-sector investment in the history of Ohio. Central Ohio has added high-tech jobs in recent years, with regional data centers being brought online by such companies as Google, Facebook and Amazon. The Intel Licking County complex is vital to both the national and economic security of the country and will be a driving force in making central Ohio the silicon heartland of the state.

BIBLIOGRAPHY

Books and Articles

Allen, Hugh. *The House of Goodyear: A Story of Rubber and of Modern Business.* Cleveland OH: Corday & Gross, 1943. Reprinted, Arno Press, 1976.

Allyn, Stanley C. *My Half Century with NCR*. New York: McGraw-Hill, 1967.

Berner, Robert. "Why P&G's Smile Is So Bright." *Business Week*, August 12, 2002, 58–60.

Borneman, Walter R. *The French and Indian War: Deciding the Fate of North America.* New York: Harper Perennial, 2007.

Boyd, Thomas Alvin. *Charles F. Kettering: A Biography*. Philadelphia: Beard Books, 2002.

Bringhurst, Bruce. *Antitrust and the Oil Monopoly: The Standard Oil Cases, 1890–1911*. New York: Greenwood Press, 1979.

Brown, Arch. *Jeep: The Unstoppable Legend.* Lincolnwood, IL: Publications International, 2001.

Building on Dreams: The Story of Honda in Ohio. Marysville, OH: Honda of America Manufacturing, 2004.

Carr, William H.A. *Up Another Notch: Institution Building at Mead.* New York: McGraw-Hill, 1989.

Clark, Henry Austin Jr. *Standard Catalogue of American Cars, 1805–1942*. 2nd ed. Iola, WI: Krause, 1985, 19.

Columbus Dispatch. "Battelle's World: Columbus-Based Research Giant Extends Its Global Reach." January 24, 2009.

Dunham, Tom. *Columbus's Industrial Communities: Olentangy, Milo-Grogan, Steelton*. Bloomington IN: AuthorHouse. 2010.

Dyer, Davis, and Kathleen McDermott. *America's Paint Company: A History of Sherwin-Williams.* Cambridge, MA.: Winthrop Group, 1991.

Evans, Harold. *They Made America from the Steam Engine to the Search Engine: Two Centuries of Innovators*. New York: Back Bay Books, 2006.

Fisher Dynamics. "History." Accessed May 25, 2022. https://www.fisherco.com/company/history.

Foster, Emily, ed. *The Ohio Frontier: An Anthology of Early Readings*. Lexington: University Press of Kentucky, 1996.

Fouche, Rayvon. "Liars and Thieves: Granville T. Woods and the Process of Invention." In *Black Inventors in the Age of Segregation,* 26–81. Baltimore, MD: Johns Hopkins University Press, 2003.

Global Security Military. "Lima Army Tank Plant (LATP)." Accessed July 6, 2022. https://www.globalsecurity.org.

Grant, Tina. *International Directory of Company Histories*. Vol. 13. Detroit, MI: St. James Press, 1996.

———. *International Directory of Company Histories*. Vol. 41. Detroit, MI: St. James Press, 2001.

Grant, Tina, and Gale Thomson, eds. *International Directory of Company Histories*. Vol. 53. Detroit, MI: St. James Press, 2003.

Hackworth, Jason. *Manufacturing Decline: How Racism and the Conservative Movement Crush the American Rust Belt.* New York: Columbia University Press, 2019.

Hall, Joseph Sparkes. *The Book of the Feet: A History of Boots and Shoes*. London: White Press, 2017.

Hart, R. Douglas. *The Ohio Frontier: Crucible of the Old Northwest, 1720–1830*. Bloomington: Indiana University Press, 1996.

Herman, Arthur. *Freedom's Forge: How American Business Produced Victory in World War II*. New York: Random House, 2012.

Historic Structures. "The Winton Motor Car Company, Cleveland Ohio." Accessed March 28, 2022. http://www.historic-structures.com.

Hobart Corporation. "Over a Century of Innovation." Accessed June 12, 2022. https://www.hobartcorp.com/about-us/history.

Hodgson, Richard S. *In Quiet Ways: George H. Mead, The Man and the Company*. Dayton, OH: Mead Corporation, 1970.

Hulbert, Archer Butler. *The Old National Road*. New York: Wentworth Press, 2019.

Ideal Electric. "An Ideal History." Accessed April 23, 2022. https://www.theidealelectric.com.

Keller, Ulrich. *The Building of the Panama Canal in Historic Photographs*. Garden City, NY: Dover, 1984.

Kepos, Paula. *International Directory of Company Histories*, Vol. 11. Detroit, MI: St. James Press, 1995.

The Kroger Story: A Century of Innovation. Cincinnati, OH: Kroger Company, 1983.

Lebhar, Godfrey M. *Chain Stores in America*. New York: Chain Store Publishing, 1963.

Lief, Alfred. *Firestone Story—History of The Firestone Tire & Rubber Co.* New York: Whittlesey House/McGraw Hill, 1951.

———. *It Floats: The Story of Procter & Gamble*, New York: Rinehart & Company, 1958.

Marion. Internet Archive Wayback Machine. June 11, 2022. https://web.archive.org/web/20080531071221/http://www.bucyru°s.com/marion.htm.

McCall, Walter P. *American Fire Engines Since 1900*. Sarasota: Crestline Publishing, 1976.

McCullough, David. *The Wright Brothers*. New York: Simon & Schuster, 2015.

Miller, Nick. "Milacron Taking on New Look Company Positioned for More Growth." *Cincinnati Post*, April 24, 1996, p. 6B.

Pederson, Jay P. *International Directory of Company Histories*. Vol. 30. New York: St. James Press, 1999.

———. *International Directory of Company Histories*. Vol. 67. New York: St. James Press, 2005.

Pound, Arthur. *The Turning Wheel: The Story of General Motors through Twenty-Five Years 1908–1933*. New York: Doubleday, Doran & Company, 1934.

Prero, Mike. "Ohio Match Factory." *Hobby History* (Nov–Dec 1997). Accessed May 19, 2022. http://www.matchpro.org.

Pruitt, Bettye H., and Jeffrey R. Yost. *Timken: From Missouri to Mars—A Century of Leadership in Manufacturing*. Boston, MA: Harvard Business School Press, 1998.

Schneider, Norris F., and Clair C. Stebbins. *Zane's Trace: The First Road in Ohio*. Zanesville, OH: Mathes Printing, 1973.

Science Serving Human Needs: A History of Battelle Memorial Institute. Columbus, OH: Battelle Memorial Institute, 1978.

Seagrave. "Refurbishment." Internet Archive Wayback Machine. Accessed June 15, 2022. https://web.archive.org/web/20070515122153/http://www.seagrave.com/content.cfm?ID=122§ionID=3.

Stover, John F. *History of the Baltimore and Ohio Railroad*. Lafayette, IN: Purdue University Press, 1987.

Sullivan, Dolores P. *William Holmes McGuffey: Schoolmaster to the Nation.* Madison NJ: Fairleigh Dickinson University Press, 1994.
Trostel, Scott D. *The Barney & Smith Car Company*. East Fletcher, OH: Cam-Tech Publishing, 1993.
Weitzman, David. *Superpower: The Making of a Steam Locomotive*. Jeffrey, NH: Godine, 1995.
Woods, Terry K. *Ohio's Grand Canal: A Brief History of the Ohio & Erie Canal.* Kent, OH: Kent State University Press, 2008.

Library Resources

Akron-Summit County Library
Ashtabula County District Library
Case Western Reserve University Library
Chillicothe and Ross County Public Library
Cincinnati & Hamilton County Library
Cleveland Public Library
Columbus Metropolitan Library
Cuyahoga County Public Library
Dayton Public Library
Huron County Community Library
Library of Congress
Licking County Library
Mahoning County Public Library
Marietta–Washington County Public Library
Oberlin College Courtright Memorial Library
Ohio History Connection
Shawnee State University Clark Memorial Library
Stark Library
State Library of Ohio
University of Cincinnati Library Archive
University of Kentucky Library Archive
University of Toledo Library
Wayne State University Library
West Virginia University Library
Wright State University Library

Newspapers

Akron (OH) Beacon Journal
Chillicothe (OH) Gazette
Cincinnati Enquirer
Cleveland Plain Dealer
Columbus (OH) Citizen Journal
Columbus (OH) Evening Dispatch
Dayton Daily News
Marietta (OH) Times
Marysville (OH) Journal-Tribune
Newark (OH) Advocate
Ohio State Journal (Columbus, OH)
Toledo (OH) Blade
Zanesville (OH) Times Recorder

ABOUT THE AUTHOR

Conrade C. Hinds is originally from Nashville, Tennessee, and a graduate of Ball State University in Muncie, Indiana, where he studied architecture and industrial technology. He has lived in central Ohio for forty-six-plus years and is a registered architect licensed in Ohio, Idaho, New York and Washington. He previously worked for the Franklin County Ohio Engineer's office and is retired as a facilities project manager with the City of Columbus Department of Public Utilities. He also served as an adjunct faculty member in the engineering department at Columbus State Community College for thirty-two years.

He has written and had three additional books published. The first book was *The Great Columbus Experiment of 1908*. The second, *Columbus and the Great Flood of 1913*, chronicles the events leading up to a record-breaking devastating flood that severely crippled industrial Midwest America. The third book is *Lost Circuses of Ohio*. It chronicles the history of the nineteenth-century circuses that rivaled Barnum & Bailey.

As a licensed auctioneer, he likes to give back to the community by donating his services to a variety of charitable fundraising events, such as Operation Feed. Conrade and his wife, Janet, live on a thirty-acre retreat in southern Ohio with their dog Porter and a cat name Fulbright. They have three adult children and five grandchildren and enjoy visiting historic sites in America and Europe. As an author, Conrade enjoys researching and writing about forgotten history.